The INHERITANCE

A FAMILY *on the* FRONT LINES *of the* BATTLE AGAINST ALZHEIMER'S DISEASE

NIKI KAPSAMBELIS

SIMON & SCHUSTER

New York London Toronto Sydney New Delhi

Simon & Schuster
1230 Avenue of the Americas
New York, NY 10020

First Simon & Schuster hardcover edition March 2017

SIMON & SCHUSTER and colophon are registered trademarks of Simon & Schuster, Inc.

For information about special discounts for bulk purchases, please contact Simon & Schuster Special Sales at 1-866-506-1949 or business@simonandschuster.com.

The Simon & Schuster Speakers Bureau can bring authors to your live event. For more information or to book an event contact the Simon & Schuster Speakers Bureau at 1-866-248-3049 or visit our website at www.simonspeakers.com.

Interior design by Ruth Lee-Mui

Manufactured in the United States of America

1 3 5 7 9 10 8 6 4 2

Library of Congress Cataloging-in-Publication Data

Names: Kapsambelis, Niki, author.
Title: The inheritance : a family on the front lines of the battle against Alzheimer's disease / Niki Kapsambelis.
Description: First Simon & Schuster hardcover edition. | New York : Simon & Schuster, 2017.
Identifiers: LCCN 2016033463| ISBN 9781451697223 (hardback) | ISBN 9781451697322 (trade paperback) | ISBN 9781451697339 (eBook)
Subjects: LCSH: Kapsambelis, Niki. | Kapsambelis, Niki--Family. | Alzheimer's disease--Patients--United States--Biography. | Alzheimer's disease--Patients--Family relationships--United States--Case studies. Alzheimer's disease--Patients--Rehabilitation--Popular works. | BISAC: BIOGRAPHY & AUTOBIOGRAPHY / Medical. | MEDICAL / Neurology. | SCIENCE / Life Sciences / Genetics & Genomics.
Classification: LCC RC523 .K37 2017 | DDC 616.8/310092 [B] --dc23
LC record available at https://lccn.loc.gov/2016033463

ISBN 978-1-4516-9722-3
ISBN 978-1-4516-9733-9 (ebook)

For HELEN KAPSAMBELIS

and GAIL DEMOE,

two women of remarkable courage,

without whom this book would

not have been possible.

Contents

The
INHERITANCE

IMMEDIATE DeMOE

GALEN "MOE"
m. GAIL HELMING

FAMILY TREE

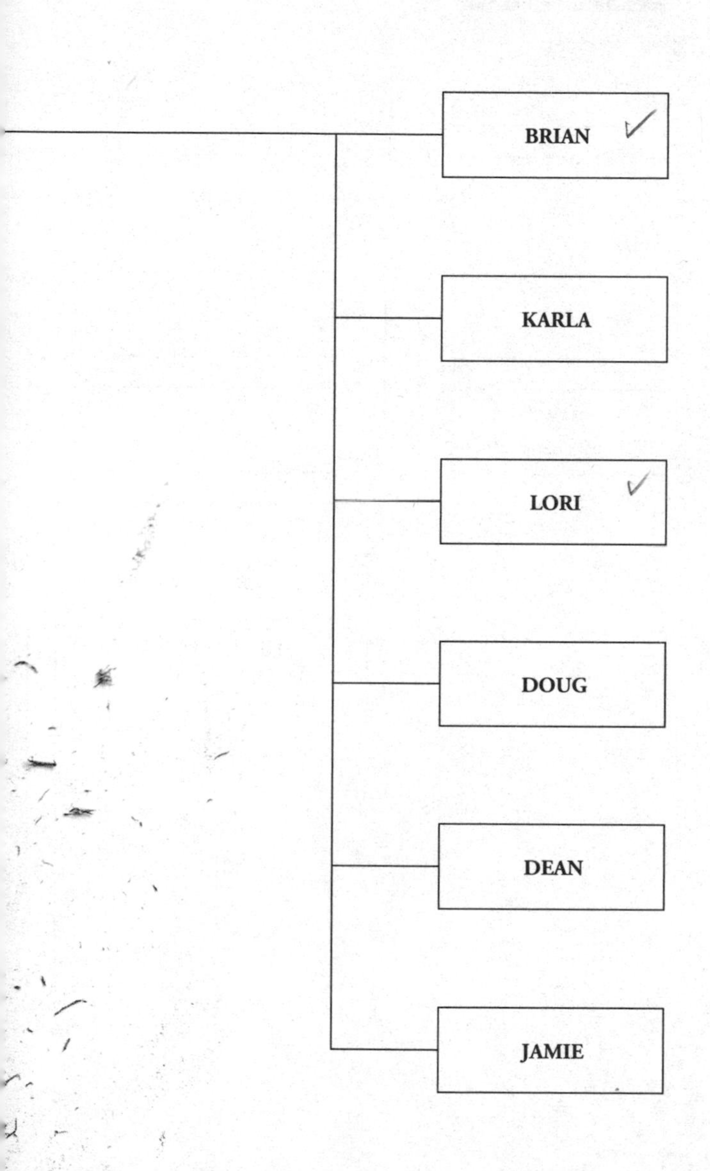

EXTENDED DeMOE

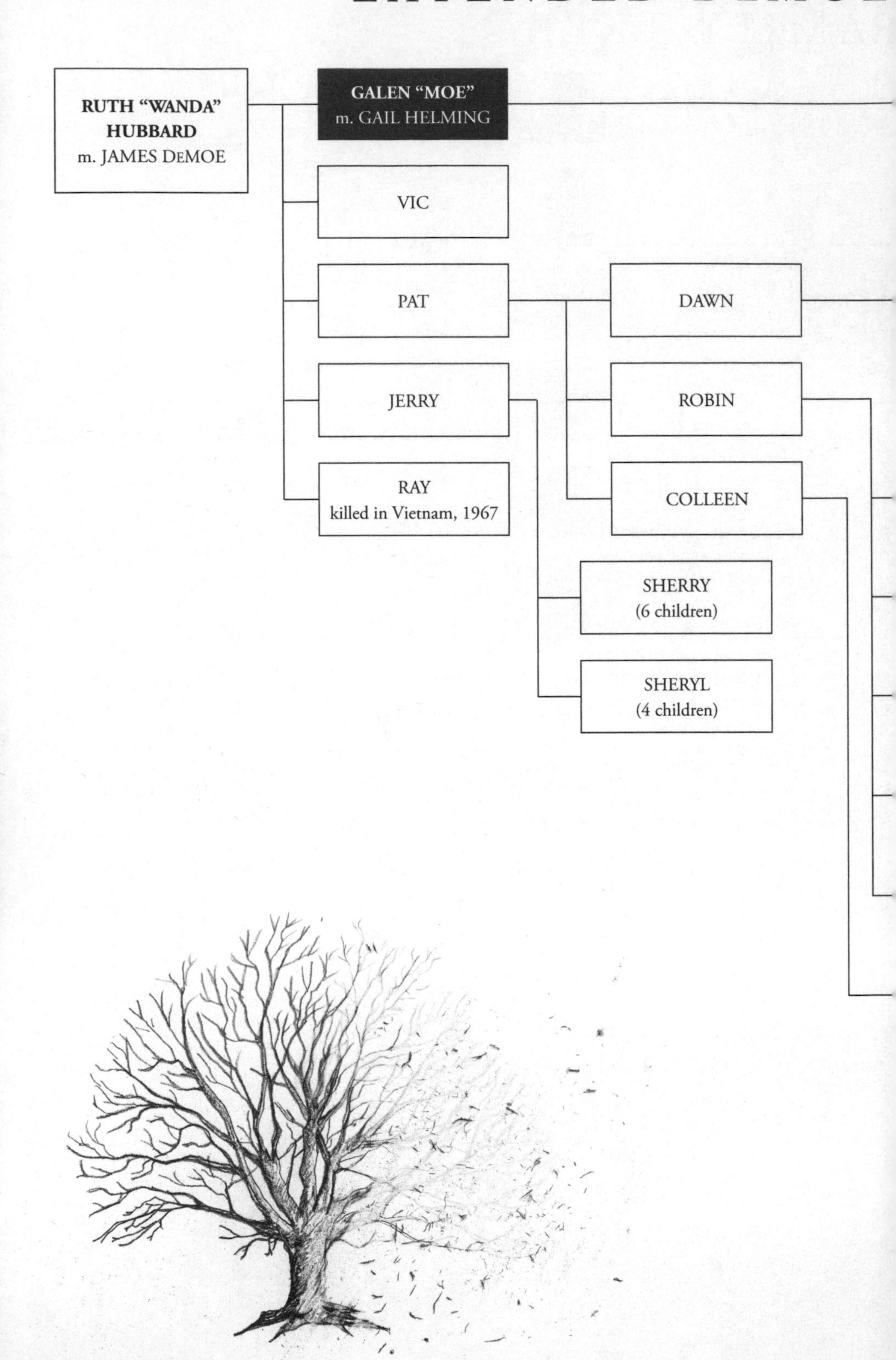

FAMILY TREE

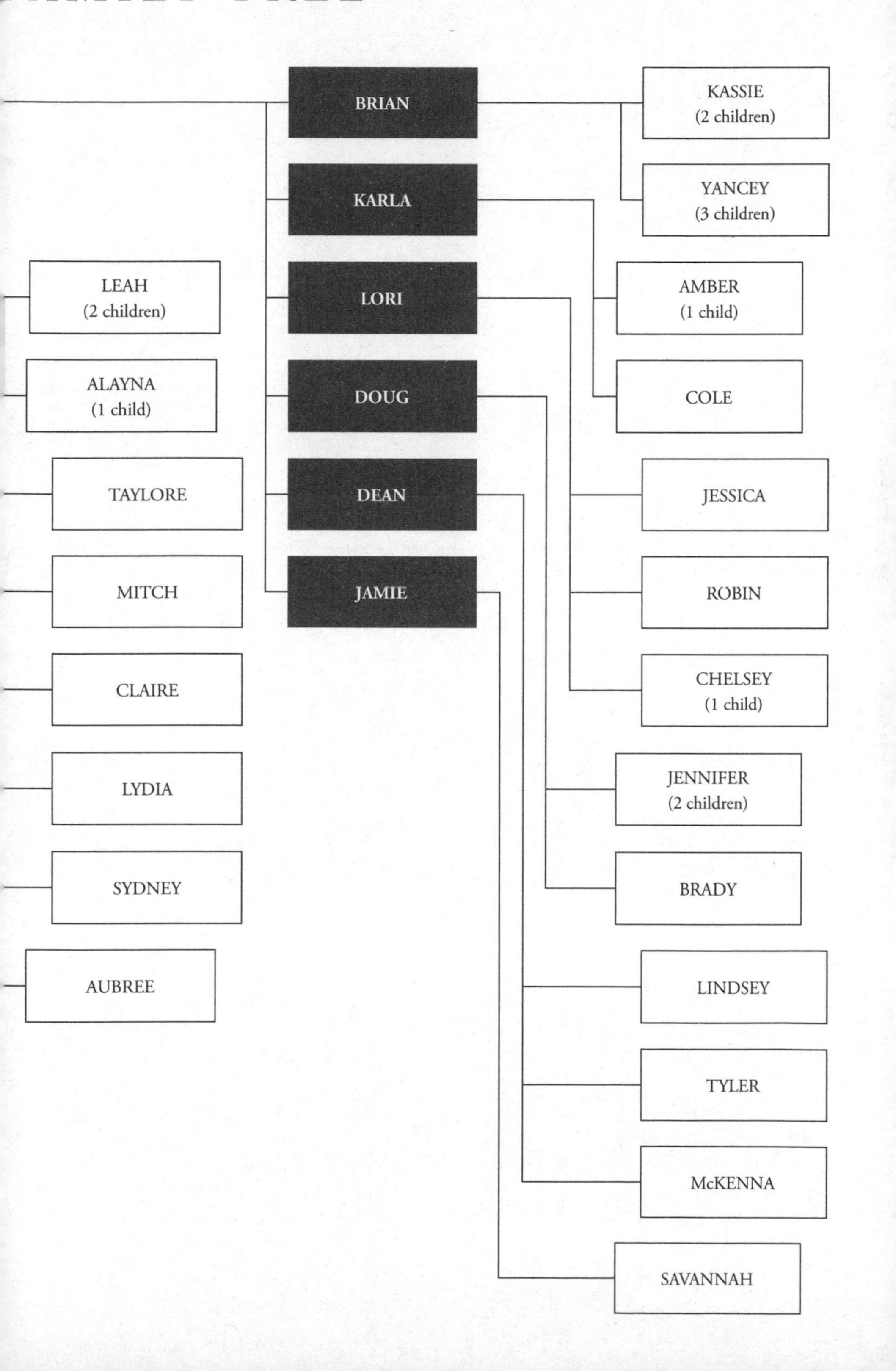

BRIAN
KARLA
LORI
DOUG
DEAN
JAMIE
KASSIE
(2 children)
YANCEY
(3 children)
AMBER
(1 child)
COLE
JESSICA
ROBIN
CHELSEY
(1 child)
JENNIFER
(2 children)
BRADY
LINDSEY
TYLER
McKENNA
SAVANNAH
LEAH
(2 children)
ALAYNA
(1 child)
TAYLORE
MITCH
CLAIRE
LYDIA
SYDNEY
AUBREE

Prologue

The doctor is businesslike, but his young face is kind. He explains that he's going to say a name and address, and he wants you to repeat it back to him, now, then once again later:

"John Brown, 42 Market Street, Chicago."

You repeat it back to him, slowly: "John Brown . . . 42 Market Street . . . Chicago."

He smiles. *Good.* He asks a series of other questions.

"And where were you born?"

"What was the last job that you had?"

"How is a turnip similar to cauliflower?"

As the questions hang in the air, the brain works to retrieve the information, stored somewhere in the hippocampus. The answers have to travel from neuron to neuron, across the spaces between the cells. They don't always arrive quickly. They aren't always complete. But they do arrive, eventually.

"What's the difference between a lie and a mistake?"

You know this: A lie is on purpose, and a mistake is an accident. The doctor seems pleased.

Then he asks where you are.

You reach for the thought, but it isn't there. When it tried to travel from neuron to neuron, it was blocked by a wad of proteins that stuck together in the space between. Without a landing place, the message evaporated, and the thought was gone.

You look around, but you don't think you've ever been here before. There are no clues to help you reach for the answer.

He asks you what day it is: You don't know. You don't know the year, or even the season. Those thoughts traveled from a healthy neuron to one that had withered up and died, and again, the message was lost.

He asks you how many nickels are in a dollar. You used to balance your checkbook with ease, but now you can't answer. It's embarrassing, because you realize it should be an easy question. You hope he doesn't tell your boss what you don't know, or your kids.

You wonder exactly who is watching this. There's someone sitting across the room, and you're not sure, but you get the impression that they're mocking you. Every time you glance their way, you catch them staring. They're even wearing the same shirt as you. You feel the anger start to rise.

He asks you again: What was that name and address?

You were telling the truth when you said your memory was OK. *A lie is on purpose; a mistake is an accident.*

You begin: "John Brown . . ."

But the rest of the answer hits another dead circuit.

In April 2009, I walked into a hotel in downtown Pittsburgh to meet a handful of members from a North Dakota family who were involved in Alzheimer's disease research at the University of Pittsburgh.

I had never been to North Dakota. In fact, I had never met anyone from North Dakota. My limited impression of the state had been formed by watching the movie *Fargo*, which is mainly about Minnesota anyway. I was equally ignorant of Alzheimer's disease—a diagnosis that, to the best of my knowledge, is absent from my immediate family tree. I walked into the hotel simply because I am a journalist, and I had been assigned to write a short article about these people.

Two hours later, I walked out, and my life had changed forever. When I reached my car, I realized I was physically shaking from the impact of their story and the magnitude of what they were doing. I immediately thought that they deserved more than the short article I had been asked to write. They deserved a book.

One day, the world will be free of this devastating disease. When future generations want to know how that became possible, may the following pages help them remember.

Early in the afternoon of a perfect July day, Dean DeMoe watched his big brother Doug fly a kite shaped like a panda.

Doug was standing at the top of a gentle slope, his fingers gently pinched around the cotton string, his arms raised skyward by the kite as it danced high in the wind that never leaves the North Dakota plains. His face beamed with the innocent, uncomplicated joy of his task; he was proud that his kite floated the highest of anyone's at the top of the hill.

Farther down the slope, close to where Dean was standing, a church group preparing for a picnic had set up a speaker that played hymns as they worked. At age fifty-two, Doug was wearing diapers. The man who had always crafted his own ice cream for family cookouts—with hands strong enough to knock an opponent out with one punch—now had his meals brought on a tray. Though he'd never been a talker, he was now unable to say more than a word or two at a time. When his language skills had almost completely left him, he broke months of silence with a single word: *Dean.*

The brothers were nineteen months apart, and though Doug was older, he had always been more of a peer than an elder to Dean. They were two sides of the same coin: Doug was taciturn; Dean was outgoing. Doug was never much of a conversationalist; he let Dean do the talking for him. Doug was a homebody, while Dean loved the adventure of travel. Dean's self-confidence was infectious, whereas Doug often gave off an air of uncertainty, as though he would rather say nothing than say the wrong thing. Dean was born to be a leader, the kind of man others would obligingly follow; Doug was the perfect employee, never missing a day of work and always glad to be given a task to complete. But on a primal level, they understood each other, their connection encoded in their DNA.

When they were children, people around town called them Deanie and Dougie. They'd been high-school wrestlers, amateur boxers, hell-raisers: the untamed DeMoe brothers. It hadn't been so very long ago. People in town still told stories about them, the hometown antiheroes. Their youth had been spent in a flurry of good-natured rebellion, punctuating long stints of backbreaking hours in the oil fields, where their father before them had once made his living. When they weren't on an oil rig, they might be found playing elaborate practical jokes or organizing parties at their family's sprawling house near the center of town. The girls had flocked to them. And they always had each other's back: anybody who crossed one brother would have to contend with the other.

Dean still put in fifteen-hour workdays in the oil fields, supervising men on a rig that could explode if somebody screwed up his job. In middle age, he was still a strong man. His crew trusted him with their lives, but he was powerless to help Doug; he could only watch as his brother slipped further away under the surface of a disease that was overtaking his brain.

As one of the church group's recorded hymns hit its crescendo, Dean turned his back to the nursing home where Doug now lived, a place Dean vowed he himself would never call home.

Dean got back in his car, slipped it into gear, and headed toward town.

Part ONE

There are only two or three human stories,

and they go on repeating themselves as fiercely

as if they had never happened before.

—*Willa Cather,* O Pioneers!

One

THE ENEMY WITHIN

WALK INTO A shopping mall. An amusement park. An auditorium of parents gathered for a school play. Within this crowd, there will be someone—in fact, several people—who are directly and irreversibly affected by Alzheimer's disease.

In the United States, Alzheimer's is the sixth-leading cause of death. Next to cancer, there is no condition more feared by human beings than Alzheimer's, for it means more than a slow death; it robs its victims of the key components of their humanity. They lose shared experiences; they fail to recognize their most cherished loved ones; they forget even their proudest accomplishments. The stress of caring for an Alzheimer's patient has decimated close-knit families, ended happy marriages, snapped the tensile bond between parents and children. And the disease is as baffling as it is unforgiving.

An estimated 24 to 36 million people worldwide—5.3 million in the United States alone—suffer from the disease or similar dementias. But Alzheimer's is the least understood of all major fatal illnesses, frequently mistaken for other conditions, especially depression if the patient is young.

Only one in four people who have the disease are actually diagnosed. None of them can be cured.

Once thought to be relatively rare, Alzheimer's is now known to be the leading cause of age-related dementia, and science is only beginning to grasp how common—and how lethal—it really is. For it is always fatal: If patients do not die from secondary causes, such as pneumonia, the disease will eventually move from erasing memory and language to shutting down involuntary functions, such as breathing and swallowing.

In the developed world, most major causes of death—including cancer, heart disease, and AIDS—have undergone great strides in treatment across the past quarter century. People do sometimes survive these diseases. But to date, science has been unable to make any kind of dent in Alzheimer's. In fact, the problem is actually growing, due to the population bubble created by aging baby boomers.

It is a disease that ignores celebrity, income, character, and gender. President Ronald Reagan had it, and so did one of his most controversial allies, British Prime Minister Margaret Thatcher. So have movie stars, literary figures, sports heroes, criminals, humanitarians, geniuses, and dullards. It has claimed victims among the most wealthy, powerful, and famous: Rita Hayworth, Norman Rockwell, E. B. White, Sugar Ray Robinson, Charlton Heston, Glen Campbell.

For such a formidable enemy, Alzheimer's managed to keep a low profile for a surprisingly long time. The disease was first identified in 1906 by its namesake, Alois Alzheimer, a German psychiatrist who was also a neuropathologist, meaning he specialized in diseases of the brain and nervous system. But descriptions of similar symptoms have appeared in literature dating back to ancient times.

In the second century, Roman emperor Marcus Aurelius employed a Greek-born physician who used the term "morosis" to describe dementia. He described people afflicted with this condition as "some in whom the knowledge of letters and other arts are totally obliterated; indeed they can't even remember their own names."

In recent years, better diagnostic tools have allowed doctors to understand two sobering facts about the way they have approached Alzheimer's disease: First, that senility is not a normal part of the aging process; people who were once generally described as "senile" often actually had Alzheimer's, meaning it is a much more widespread disease than anyone realized. And the second fact, which is more frightening, is that no current medical intervention can reverse it, or even slow it down, because for most of the time science has known about Alzheimer's, there has been no way to see it coming until it has already wreaked havoc within the walls of the brain.

From 1906, when Alois Alzheimer first described the disease, until well into the twenty-first century, diagnosing Alzheimer's disease in living patients was little more than an educated guess. Doctors relied on clinical tests, asking questions about the patient's memory and ability to function. Though these tests depended on the patient's honesty, doctors might separately verify answers with close friends or family members. There really weren't objective physical tests, although there were some telltale physical signs, such as a shuffling walk. Mood changes could occur, too; aggression, hallucinations, and depression were common.

But all of these symptoms can also point to other afflictions: meningitis, brain trauma, stroke, syphilis, and medication side effects can produce similar results. Even sleep apnea and urinary tract infections can cause confusion. And while Alzheimer's disease is the leading cause of dementia, accounting for 60 to 80 percent of all cases, other causes exist, too, such as Parkinson's and Huntington's diseases. The word "dementia" is a general catch-all term encompassing many abnormalities.

A study of 852 men diagnosed with Alzheimer's disease from 1991 through 2012 found that the diagnosis was wrong one-third of the time, correct one-third of the time, and partially wrong—in other words, the patient had a mixture of diseases—one-third of the time. And in situations where a patient is young or a doctor has limited experience with memory disorders, the diagnosis becomes even more elusive.

For most of the time science has known about the disease, a true, definitive diagnosis of Alzheimer's—not *probable* Alzheimer's—could only happen after death, when a neuropathologist examined brain samples under a microscope to confirm the presence of amyloid plaques and tau tangles, the abnormal proteins that are the disease's grim signature. Plaques are sticky, microscopic clumps of stray amyloid proteins that form outside the brain cells and possibly prevent the cells from signaling each other. Tangles occur inside the brain cell. They are twisted fibers of the tau protein, which—in its normal state—helps transport nutrients. When its strands begin to twist, they choke the transport system and the cell dies.

Current consensus within the Alzheimer's research field holds that early intervention is key; by the time a person shows what we think of as mild symptoms, such as occasional forgetfulness, the brain may have reached a tipping point from which it will not return. But just how far in advance a doctor would need to give a treatment isn't known. Is ten years before the onset of symptoms soon enough? Should it be sooner? Can it be later? If scientists were working with a patient who knew that he would develop Alzheimer's at a specific age, they could answer these questions faster.

So even as they search for a viable treatment, researchers also continue to seek out ways to predict who the disease will strike. If they know who will someday get Alzheimer's, they want to treat that person before he begins to slip away, much the way possible cardiac patients are now given cholesterol-lowering medication to help them avoid heart attacks. But to find such a treatment, doctors need a patient who is guaranteed, with 100 percent certainty, to get the disease—only then will they know if an experimental treatment was successful, by testing it out on that person and then measuring its effect.

Those perfect patients do exist, as one tiny sliver of the population who stand distinctly apart from the rest. They are the people living with one of three known genetic mutations that guarantee they will be stricken. Only about 1 percent of all Alzheimer's patients fall into this category. They are hit young: Their average age of onset is between thirty and fifty years old. Often, they have children, not knowing they stand a 50 percent chance of passing on the mutation; so the disease has raged silently through generations of families. For as little as science has known about Alzheimer's, it's known even less about these mutations.

But in nature, curses are often a double-edged sword. As tragic as mutations are, they may well hold the key to preventing—or at least delaying—Alzheimer's. Doctors can diagnose patients with mutations years before symptoms appear, even in childhood. By testing preventative drugs in this population, researchers hope—and the rest of the world prays—that they will be able to translate a successful treatment to the rest of humanity before another generation is lost.

To get to that point, quiet sacrifices have been made by the most ordinary of people. They could be your neighbor, your coworker, your high-school classmate. Their lives were sometimes colorful, sometimes simple; but in their mutations, they have become exceptional. For it is their courage, often driven by desperation—sometimes tempered by fear or frustration—that has fueled the science that hopes to beget the solution. These are the people future generations will thank when Alzheimer's itself becomes a distant memory.

Alois Alzheimer was a bespectacled, cigar-loving, robustly built man. He took a job in 1888 at the Frankfurt Asylum for the Insane and the Epileptic—a facility housed in a fairy-tale Gothic revival building known colloquially as the "Castle of the Insane."

A few years before Alzheimer joined the staff, the assistant medical director of the Frankfurt asylum, Franz Nissl, had invented a method for staining brain cells, turning their components a vivid shade of methylene blue that made them easier for researchers to analyze. The process, known simply as the Nissl stain, is still in use today.

In Nissl, Alzheimer found a lifelong friend. They shared a professional interest in linking symptoms of mental illness to physical causes through a microscopic analysis of the brain. Better imaging, they reasoned, would allow doctors to more clearly define and treat the disorders. In those early years, Nissl and Alzheimer worked as clinicians by day and conducted their research by night, frequenting pubs when time permitted. Nissl served as a witness when Alzheimer married his wife, Cecilie, the widow of a wealthy diamond dealer who had briefly been his patient. But in 1901, Alzheimer suffered a devastating personal loss when Cecilie died months after giving birth to their third child. Grief-stricken, he buried himself in his work, personally seeing virtually all newly admitted patients and committing his findings to an extensive written record. Cecilie's fortune would later allow Alzheimer to devote all his time to research, a rarity for that era.

The same year that Alzheimer lost his wife, a Frankfurt railroad worker named Karl Deter was also losing his.

Auguste Deter was a wife and a mother, hardworking and orderly. In school, she may have been a student of Alzheimer's grandfather, Johann. She married her husband, Karl, in 1873, and together they had a daughter, Thekla.

In March 1901, just before her fifty-first birthday, Auguste began developing the bizarre symptoms that would mark her rapid decline into dementia. Although she'd always had a somewhat excitable personality, she became inexplicably and irretrievably jealous, accusing Karl of having an affair with their neighbor. She began blundering through the cooking and the laundry, and she started squirreling objects away in their home. She cried constantly. She was convinced that a courier who frequently stopped by was plotting to hurt her.

"She lives in a world of the moon," Karl Deter reportedly said to a work colleague. "Even my jackets are badly cared for."

Things continued to roll downhill in the Deters' home. Unable to sleep, Auguste sometimes wandered at night, or worse, woke up screaming uncontrollably. She deteriorated to the point where she could not handle any type of work. She busied herself with plans to visit her mother, who had been dead for more than ten years. She accused her husband of hiding jewelry she had inherited from her grandmother.

Auguste was admitted to the Castle of the Insane, where Alzheimer took notes on his first visit with the new patient on November 26, 1901:

She sits on the bed with a helpless expression.

What is your name? Auguste.

Last name? Auguste.
What is your husband's name?
Auguste, I think.

All told, Auguste spent close to five years at the asylum, and by any account, it was hard time. Moody and anxious, she alternated between calling for her husband and daughter and failing to remember parts of her own name. Sometimes she withdrew or whined; she continued to hoard objects, this time under her bed, and dragged the bedclothes around or buried herself under them. She was not allowed to wander freely because she would become aggressive with other patients and grab their faces, although sometimes she was also kind and courteous. Most of her days were spent in the bathtub, a common remedy that was intended to soothe agitated patients. Her nights were spent in the ward's isolation room. "I have lost myself," she confided to her doctor.

Although Alzheimer had seen patients with similar cognitive deterioration, most of them had been much older than Auguste Deter—in their seventies, not early fifties. He attributed their dementia to atherosclerosis, a thickening of the brain's blood vessels. He continued to study his unusual and otherwise healthy patient, calling her malady "the disease of forgetfulness," never realizing that her affliction and those of the older patients were likely one and the same.

"Are you sad?" Alzheimer asked her on a visit in early December 1901.

"Oh always, mostly not," she answered. "It happens that one sometimes has courage."

In 1902, Alzheimer left the asylum to take a new job working with the most respected German psychiatrist of the day, Emil Kraepelin. He hadn't been able to cure or even successfully diagnose Auguste, but his fascination with the weeping, lost woman never waned; he kept track of her condition with the help of his former boss. Auguste's husband, who struggled to pay her medical bills, also visited her when he could. By 1905, she was bedridden and incontinent, unable to feed herself. Her weight dropped to sixty-eight pounds, and she lay curled in the fetal position. Her agitation stopped responding to sedatives. A bedsore festered into sepsis and pneumonia, and she died on the morning of April 8, 1906, a month shy of her fifty-sixth birthday.

Twenty days later, Alzheimer's former clinical director at the Castle of the Insane sent a box containing Auguste's brain, brainstem, spinal cord, and medical records on a train to him in Munich, 190 miles away, where Alzheimer spent the next six months analyzing her disease. At his disposal was a laboratory outfitted with the most modern equipment available, including the first distortion-free microscopes.

Alzheimer's assistants prepared more than 250 slices of Auguste's brain and spinal cord into slides stained with several different techniques, including the one invented by his old friend Nissl, to help him better examine the intricacies of the cells. There, Alzheimer got his first glimpse of Auguste Deter's enemy within.

As he studied her cortex, the brain's largest section and the one that controls higher functions such as thought and action, he saw that it had been taken hostage by brown clumps of plaque, sticky blotches resembling tumbleweeds that had landed in the space between neurons. A different stain revealed all manner of tangled fibrils—dark, twisted bundles resembling balls of twine, crescents, and baskets, growing out of control and wiping out a third of the neurons in her cortex. In short, Auguste's brain, like her body, had atrophied, apparently thanks to normal cell components that had somehow turned traitorous.

With so much cellular death, Alzheimer felt certain the lesions had to be the key to Auguste's bizarre behavior. What surprised him most was how extensively the brain had changed—more than people in their seventies and eighties, who typically experience a loss in brain volume as part of the aging process—even though she was just fifty-six. It was not like any illness he had ever seen before. Excited by his discovery of what seemed to be a new disease, Alzheimer carefully prepared a lecture on his findings for the 37th Assembly of Southwest German Psychiatrists in Tübingen. Eighty-eight respected colleagues were in attendance, including Nissl and the child psychologist C. G. Jung. When Alzheimer concluded his remarks, he expected an avalanche of questions.

Instead, he was met with deafening silence. The conference chairman, who was acting as moderator, repeated himself: Did anyone have a response? They did not. He thanked Alzheimer for his presentation and moved on.

In hindsight, it's difficult to imagine a lack of interest in the discovery of such a widespread disease. Yet within the context of early twentieth-century psychiatry, the collective shrug was not surprising. The field generally did not believe in a correlation between mental illness and biological causes. (One notable exception was general paresis of the insane, a form of dementia caused by syphilis.)

Alzheimer hoped his discovery might help underscore the connection between brain and behavior. But Auguste Deter's case seemed too rare to be of clinical importance, given that no link had been established between her disease and the more common senility of older patients.

The audience was eager to move on to the presentations that followed, which delved into the sexier issues of hypnosis, childhood trauma, and Sigmund Freud's new field of psychoanalysis. Disappointed, Alzheimer packed up his slides and left the stage. The local newspaper devoted one sentence to his talk.

The disinterest of the psychiatrists at the conference would prove an unfortunate foreshadowing for the way the field treated Alzheimer's disease for the next several decades. Scientists simply didn't understand that they were dealing with a disease that was affecting people all around them; it would remain an invisible predator—its power unchallenged.

Although Alzheimer's presentation flopped at the conference, he retained the enthusiastic support of his boss, Emil Kraepelin, who shared the radical theory that mental disorders had physical causes and was pleased to promote the discovery. Four years after the Tübingen conference, Kraepelin coined the term "Alzheimer's disease" in the eighth edition of his book *Psychiatrie*, or *Handbook of Psychiatry*.

Ironically, Kraepelin—who valued classification—unwittingly worsened a key confusion over Alzheimer's disease by defining it as a dementia that occurred in patients before the age of sixty-five. After that arbitrary milestone, it was the much more common senile dementia, he said. And senility was so common that it was thought to be a standard part of aging, like graying hair or sagging skin.

Alzheimer himself didn't dispute this. In 1911, he wrote that while the two diseases were similar, he could not be certain they were identical. Yet the only difference was that the dementia happened to some people earlier than others. This misperception would muddy the waters for decades, allowing this widespread disease to go largely unexplored.

Unfortunately, neither man would live to see the discovery vindicated. During the next fifty years, little public fanfare was given to Alzheimer's findings, which were thought to be interesting but too rare to be of larger significance. Plagued by heart and kidney problems, Alois Alzheimer died in late 1915. The slides he made from slices of Auguste's brain, as well as his notes, clinical materials, and case histories in both Latin and German, were added to blue cardboard files and left to collect dust deep within Johann Wolfgang Goethe Frankfurt University Hospital, where they lay until 1995.

Two

THE SALT OF THE EARTH

IN 1973, FORTY-THREE-YEAR-OLD Galen DeMoe sat in his pickup, surrounded on all sides by the North Dakota oil fields where he'd worked his entire adult life, the rigs pumping oil out of the dirt while crews alternated between backbreaking, treacherous labor and long stretches of boredom.

To an outsider, the rigs can all look the same, metal towers whose machinery moves day and night, the ground underneath spattered with petroleum. But a practiced oilman can see in an instant which rigs are well maintained, which ones will feed his family, which ones are an explosion waiting to happen. Galen, whom everyone called Moe, had been working in the oil fields for a quarter of a century and knew them like the back of his weather-beaten hand.

But on this day, he sat in his truck, confused and desperate, with no idea where he was supposed to go.

Impossibly, when a coworker found him that way, Moe was crying.

• • •

Galen DeMoe was a larger-than-life figure in the small North Dakota town of Tioga. He was a hard drinker, a harder worker, and one of the best-liked men in the oil fields. Nobody could outpace Moe when it came to his work ethic. He would drive through the most brutal ice storms to get to a job, put in shifts that would have broken a lazier man, and he'd still make time for a shot and a beer afterward, recounting the day's war stories with a laugh. He never left his coworkers hanging out to dry, and on an oil rig, that was a virtue that could save lives. Although he harbored a soft spot for children, at home his temper emerged, which his four sons and two daughters had learned to fear, his face screwed up in anger beneath his perpetual flat-top haircut. Nevertheless, their household was always full of laughter and love, due in large part to his charismatic and devoted wife, Gail.

> *That was a good picture of you just as good looking as ever. Be good now and let those boy[s] alone in school.*
>
> *—excerpt from a letter written by Moe to Gail, April 17, 1951*

Born in 1930, Moe was raised on a farm in west-central Wisconsin, the oldest of five. He and his siblings were never especially close, perhaps in part because of a wide age gap: The two oldest children, Moe and Vic, were three years apart in age; but the third child didn't arrive until Moe was fourteen, with two more born soon after that.

Seven hundred miles away, in North Dakota, young Gail Helming was a lively girl, creative and witty, with a delicious sense of humor. Small-framed, with deep dimples and wavy brown hair, she wore cat's-eye glasses and the prim fashions of the early 1950s—Peter Pan collars and high-button blouses. Gail maintained an ongoing love affair with the written word; she composed poetry, played the piano, and on occasion would spontaneously break into song. Popular and high-spirited, she loved to make people laugh. Nobody would ever accuse Gail of being an introvert.

She may have made friends easily, but Moe was the first and only man Gail ever loved. In 1947, when he was sixteen years old, Galen DeMoe left his family's farm just outside Eau Claire, Wisconsin, and hitchhiked west with a friend to seek his fortune in the oil fields of North Dakota, where the two boys boarded with Gail's mother.

For twelve-year-old Gail Helming, it was love at first sight: "It's funny how at that age, sparks can fly. Me, I thought he was the best thing since 7UP," she said. She saw fate in their similar names: Clearly, Galen and Gail were meant to be together.

Their romance would build slowly over the next decade, through his four years in the military and a stint working in Alaska. Moe came home from the frontier as soon as he had enough money for an engagement ring and a car.

Gail didn't know much about the family she was marrying into, a fact that would haunt her for the rest of her life. She traveled to Wisconsin to meet Moe's parents, where on that visit Wanda DeMoe did act a little strange. When Gail arrived, Wanda was preparing to host members of her local Homemakers Club, and she spent Gail's visit cleaning and putting out desserts instead of getting to know the girl who had won Moe's heart. Gail wondered why Wanda hadn't simply told her this would be an inconvenient time to arrive. Wanda's housekeeping seemed disordered, too, but since her future mother-in-law wasn't physically strong, and she was going through menopause, Gail just dismissed her domestic quirks. She never minded a little eccentricity; in fact, she embraced it. "I was in love, no matter what," she said.

Though Gail was just twenty, many of her friends were already married. Eight years after she first laid eyes on him, she and Moe became engaged in January 1955, and they wed the following August.

"We had the reception at my folks' house, which is unusual," Gail recalled. "And the priest got drunk. He was supposed to go and hear confessions on the other side of town, but he didn't make it."

The newlyweds settled in Tioga, their arrival coinciding with one of the town's many oil booms, meaning Moe could expect steady employment. The self-proclaimed "oil capital of North Dakota," Tioga lies nestled in the state's northwest corner, about an hour's drive south of Saskatchewan or east of Montana. It was founded in 1902 by homesteaders from Tioga County,

New York, and the presence of the Great Northern Railroad meant that from its inception, it was a town that attracted outsiders seeking their fortunes.

On April 4, 1951—four years before Gail and Moe were married—the Amerada Petroleum Corporation discovered oil in Tioga on the Clarence Iverson farm. Around the same time, wells were drilled on the Henry O. Bakken farm northeast of Tioga, which was the first recorded extraction of oil from what is now called the Bakken Formation—approximately two hundred thousand square miles of oil-rich rock stretching underneath North Dakota, eastern Montana, and southern Saskatchewan.

Those discoveries sparked Tioga's first oil boom, setting off a chain reaction that echoed loudly into the next century. The US Geological Survey estimates that 4.3 billion barrels of oil lie in the Bakken, though some peg its rocky reservoirs as high as 200 billion.

With the oil boom, the population burst wide open, fueled by a steady influx of speculators, oil companies, and laborers hoping to benefit from the discovery of black gold on the North Dakota plains. Tioga grew from about 500 people to 2,700 in 1959.

The newly minted Mr. and Mrs. DeMoe settled into an eight-foot-by-thirty-three-foot trailer provided by Halliburton for its employees, and Moe set to work in the oil fields.

"Then the family started," said Gail. "Once I got the machine going, I couldn't get it turned off." Nor did she want to, because Gail was a born nurturer. People were drawn to her natural sense of empathy and her willingness to accept them.

Brian came first, in 1956, and by everyone's account, he was his mother's favorite. He was a miniature Moe, all freckles and swagger. Karla, Lori, Doug, and Dean followed in quick succession. The youngest, Jamie, arrived in 1971, nine years after Dean. The other children had hoped Jamie would be a girl, so they could be like the family on *The Brady Bunch*: three boys and three girls. Though they never achieved that symmetry, the siblings' bonds were strong.

As the DeMoes grew, they moved from their tiny trailer and found a clay-colored rambler at the corner of North Hanson and Second Streets that was just right for a large, boisterous family. It was a one-story house with the biggest living room they'd ever seen and a galley-style kitchen that was never quite big enough.

It was a noisy, riotous house that resembled Grand Central Terminal more than the split-level perfection of *The Brady Bunch*. Every corner, every cupboard was steeped in the personality of the people who lived there. From the battered upright piano in the living room came the sound of students plunking away at the lessons Gail gave; in the refrigerator lurked jars and containers of sauces and contents of mysterious origin. Everyone was too busy to throw them out. In an unused, detached garage out back, the boys snuck beers, risking their father's wrath. Late at night, their friends tiptoed down into the finished basement so frequently that thirty years later, they still remembered which squeaky steps to avoid.

The DeMoe house was destined to become a town institution, with petite, indomitable Gail at its core. After hours, when the bars closed, it was not unusual to find customers drifting down the street and through the large corner lot into the unlocked back door, where they'd fix themselves snacks in her kitchen. Gail DeMoe never minded. Her sense of humor was notorious. When her son Brian chased her through the house, spraying her with a garden hose, she laughed hysterically. Once, she popped out of a giant cake to celebrate the birthday of a friend who was expecting a show-girl. She briefly worked as an entertainer at local children's parties, and kept a collection of costumes and wigs in her basement for kids to use for dress-up, frequently joining them. She would become the town's most popular citizen, with a sometimes ribald sense of humor that immediately made people feel at home. Her favorite saying, which she often repeated to her children and friends, was: "Love you to the moon and back."

Gail formed a unique bond with each of her offspring, following the contours of each personality as if that one were her only child. She allowed Doug his lengthy silences, and he gratefully returned to the comfortable acceptance of her kitchen throughout his adult life. She snuck cigarettes with Dean, her rebel, and laughed off Karla's criticism of her decorating style, which tended toward knickknacks that Karla found tacky. Even Jamie, who defied her so sharply in his preteen years that she sent him to live with Karla for a brief time, mowed her lawn for her and sought her advice. Several of her sons' girlfriends viewed Gail as a confidante, and she sewed a prom dress for Brian's date that matched his shirt perfectly.

To support the warm household that Gail was building, Moe worked punishingly long hours, as would his sons when they eventually joined him in the oil fields. Over the years, the job would cost him the tip of his left index finger and two toes on his right foot, but he never quit, or even slowed down. When he started, he drove a cement truck, then later he moved into a job testing drill stems. It was a dangerous, demanding industry, but a hard worker could earn enough pay to support six children when the oil was flowing. More than the paycheck, Moe's identity was wrapped up in his work ethic. It was his chief virtue, a defining characteristic in which he took enormous pride. He was frequently gone both day and night, even when the kids were little. If his temper was short when he finally came home to the chaos of so many children, nobody could really blame him.

"As far as Moe goes, he was one of the better guys to work with," said Hank Lautenschlager, who lived across the street and joined the Halliburton crew. "He was so well liked by everybody."

Handsome and inseparable, Dean and Doug—the fourth and fifth DeMoe children—shared a love of wrestling, boxing, and football. Though Doug was older, he was the quieter of the two boys, his introverted personality overshadowed by Dean's sociable good nature. Dean loved to drive his car too fast, to strike up conversations with pretty girls, to tell funny stories from work about the idiot who rode down a cable wire from the top of the oil rig to the ground.

Doug preferred to let his actions speak for him. At the town pool, instead of drawing a crowd by telling anecdotes, Doug—a strong swimmer—would turn backflips into the water to attract attention. He loved water-skiing behind a speedboat on the lake formed by the Tioga Dam. When he developed an interest in motorcycles, he showed off by executing perfect wheelies that he could hold for as long as a mile. He and his best friend, Gary Anderson, sometimes picked up what passed for town derelicts in Tioga—"dope heads," Gary called them—and talked them into going on a road trip, then drove them a hundred miles west and stranded the poor sots in Montana. They figured they were doing the town a favor.

The oil fields taught the boys that hardships and dangers were unavoidable in life, and that they needed to grasp and savor what was good. Dean, in particular, was beginning to prove himself as a man by cultivating a reputation as a risk taker: There was no dare he would not accept, and people loved him for it. But his daredevil façade hid his sensitivity. He would grow into a thoughtful man, protective of those he loved, perhaps striving to show them the warmth he never felt from Moe.

In ways he probably did not recognize in himself, Doug inherited many of his father's personality traits. His best friend, Gary, perceived that they were both, in their own way, very soft-spoken—even though Moe had a fearsome reputation among his offspring, Gary thought of him as a kind man in the burly body of an old-fashioned wrestler. Like Moe, Doug's sense of self-worth hinged on his job status. He wasn't a particularly creative thinker, the way Dean was, but he was a completely reliable worker, never taking a day off if he could possibly help it. And like Moe, he had a soft spot for small children.

Where Doug and Dean's personality differences drew them together, Karla and Lori's created more distance. Both were cheerleaders, but that was just about the only interest they had in common. Though they were a year apart in age, they wouldn't share much of an emotional bond until many years later, when they were mothers themselves.

Karla, the second oldest, was a typically self-centered teenager, worried about appearances and anxious to fit in. Her family's popularity was important to her, a kind of status symbol. She liked being one of them, a big fish in Tioga's small pond. The familiarity of her world was comforting, and she was easily flustered when she moved too far beyond it. She relied on her family and friends for guidance, secure in the knowledge that being in a large family meant always having backup when life got a little too scary.

Just as Brian, the oldest, was their mother's favorite, Karla was Moe's, the only child who didn't get the brunt of his famous temper. She was traditional and family-centric, a homebody who met her future husband at an eighth-grade dance, then stayed with him for the next forty years. Often, her chipper personality hid a lack of confidence. She got lost easily, for example, even when driving roads she had known her entire life.

But Lori, the third child, was a free spirit. As much as Karla wanted to fit in, Lori loved to stand out. She was a wild rose, never caring what anyone thought of her, and she found Tioga stifling. She longed for new experiences, new faces, to explore the unknown.

"She drove this old car that I wouldn't be caught dead in, and wore these long earrings that I wouldn't be caught dead in," said Karla. Back then, those things mattered in a way that seems impossible now. They had their own friends, and even in a small town, their social circles didn't mix.

Lori wore her brown hair boyishly short and worked in the parts department of a local car dealership. She joined the cheerleading squad only so she could take the team bus to other towns and scope out boys she hadn't known since kindergarten. Her rebellious streak rivaled Dean's, but she was also a nurturer, like their mother. If a dog was ugly, she loved it even more because it seemed to need her. She connected easily with people on society's fringes—the overlooked, the castaways—and she rejected anything pretentious. Most important, when Lori had a problem, she went looking for her own solutions, and prided herself in her fierce independence. She did not like answering to anyone.

Among the siblings, no quarter was given for sensitivity; all character flaws—Doug's stingy, tight-lipped silences; Jamie's perpetual worrying; Karla's prissiness—were targets for teasing. The constant jokes created a camaraderie that drew them closer together, because they accepted one another unconditionally, flaws and all. The teasing also toughened them up. While they did argue, they didn't go for months without speaking; Gail would not allow it. She set the example for them, always keeping in touch with her own siblings and friends, reminding them how important they were to her, even when they were at odds.

The jokes also served as a handy way of deflecting attention from their father's strictness. They didn't want pity. They wanted people to laugh with them, not at them, and they wanted everyone they knew to be in on the joke—one of the keys to their popularity.

Their love for pranks sometimes bordered on public nuisance. The boys took potshots at a statue in the center of town and shot out streetlights. One of their favorite practical jokes was "canning cars." The DeMoe boys threaded beer cans along a very long strand of fishing line, which was then stretched across the street. When unsuspecting cars whizzed by, boys on either side of the street would raise the line just high enough to catch in the car's grille, and the driver would continue on, dragging a clattering tail of aluminum empties behind him. Even the Tioga cops knew about the can trick; they chuckled and turned the other way. One officer, who also happened to be Doug's boxing coach, sometimes pulled the boys over when he saw them out driving around, then joined them in the car for a beer before resuming his shift. Tioga's culture was solidly rooted in its wild frontier past.

Laughter infused the family through its early years, embedding itself as their foundation hardened, ready to be extracted when times became impossibly hard.

That winter day when an oil hand found Moe crying in his truck between job sites was the first time anyone outside of the house had noticed his trouble. Gail herself had seen other changes, but kept them to herself.

It started with the smallest things. Moe had a hard time remembering phone numbers, so he scribbled them on the back of his hand. When he wanted a cup of coffee, he struggled to remember how to turn on the pot. The man who had loved sports since childhood abruptly stopped playing softball with his friends. Moe no longer made time for a round on Tioga's golf course. Gail watched and wondered, but at the time it didn't occur to her that something might be starting to go drastically, terribly wrong.

"Dummy me," she said later.

Even more obvious, in retrospect, was another warning sign: The same thing had happened to his mother. Gail and Moe's children had been in elementary school when Wanda DeMoe, back home in Wisconsin, began to seriously decline. Her two oldest children had already moved out: Moe was in North Dakota, and Vic, a former Marine, was working as a policeman after returning home from Korea. But Wanda's three youngest children were still home. The next oldest, Pat—the only girl—was barely in her teens when her mother began to forget; the little boys, Jerry and Ray, were even younger.

"It was really hard for the kids," Gail remembered.

Nobody could really say what was wrong with Wanda. Doctors originally theorized that it was multiple sclerosis, but this was the late 1950s; it was a murky diagnosis, at best.

Wanda was in her forties when her symptoms surfaced. To avoid parallel parking, she instead usually stopped across the river that ran through her town, leaving her grade school–aged children in the car as she walked across the bridge. When she was finished with her errands, she took the bus home, completely forgetting about the kids or the car. She couldn't be trusted alone; she left food cooking on the stove while she was out of the house. And when she walked, her steps turned into shuffling, like a woman twice her age.

Whatever was ailing Wanda also seemed to be affecting her two brothers. At one family gathering, both men saw their reflections in a glass door and demanded to know who those strangers were.

When Wanda's husband took her to a doctor, she was prescribed shock treatments to address a possible nervous breakdown. Pat, her young daughter, took over most of the housekeeping. Eventually, in her early fifties, as her speech became unintelligible, Wanda went to live in a nursing home. She never had a true diagnosis while she was alive.

Gail went to visit her mother-in-law and was horrified to find Wanda in a facility full of moaning, screaming people. To her, the ward looked like something out of *One Flew Over the Cuckoo's Nest*. Nobody was physically active; most people seemed to be bedridden. Seemingly overnight, Wanda's hair had turned white, and she lay curled in the fetal position.

On July 15, 1964, five years after her strange symptoms began, Wanda DeMoe died at the age of fifty-four. Bedsores covered her hips, buttocks, and shoulders. The doctor who performed her autopsy peered into her brain and found that her cortex had, like Auguste Deter's, atrophied and was speckled with dense clusters of tangles.

Wanda DeMoe did not, as it turns out, have multiple sclerosis. Wanda DeMoe had Alzheimer's disease.

About twenty years prior to Wanda's death, science began establishing that at least some forms of Alzheimer's might be genetic. And in 1963, the year before she died, a group of doctors led by Robert Feldman, then of Harvard University, wrote about members of a family who seemed to be inheriting the disease in a pattern that suggested it could be passed down by one parent, known in genetics as autosomal dominance. But it would take another decade before researchers would more fully understand that pattern as it applied to Alzheimer's.

Wanda's younger children dispersed not long afterward, their father apparently unable to take care of them on his own. Pat got pregnant at sixteen and married her boyfriend; Jerry, the second youngest, went to live with Moe and Gail in North Dakota, where he finished his last five years of school.

Jerry adapted quickly to his oldest brother's household, agreeing to babysit his nieces and nephews—who were not much younger than he—so Gail and Moe could go out for the evening. Then he shooed the children into their rooms so he could throw parties. The DeMoe siblings loved to peek in on the fun, and Jerry's friends bribed them with amusing stories and candy if they didn't tattle.

Kindhearted neighbors took in Moe's youngest brother, Ray, who had grown close to them while working on their farm and didn't want to relocate out of town. He planned to become a farmer one day, but first he followed in his brothers' military footsteps, joining the Marines during his senior year of high school. In November that year, he was shipped to Vietnam.

On March 25, 1967, Ray's platoon was pinned down by enemy fire. They were running low on ammunition, so Ray, heedless of his own safety, tried to get some. Exposed, he was killed by the mortar fire. Private First

Class Raymond DeMoe was nineteen when he died. There was no way to know if he had inherited the same illness that had killed his mother.

Nine years after Wanda's death, Gail's growing concerns about Moe prompted her to take action. In 1973, they visited a neurologist in Fargo who ran him through a battery of paper-and-pencil tests, took X-rays, and attached electrodes to his skull for a neuroencephalogram, which Moe blamed for giving him a terrific headache.

The doctor called Gail into a room by herself to discuss his findings: He thought Moe had dementia, and that it could be hereditary—a remarkably perceptive diagnosis, especially for the time. It was then that Gail first heard the word "Alzheimer's" used to describe her husband's condition. She had never heard of it. Seven decades after Alois Alzheimer's discovery, it was still considered a rare affliction, though the doctor who had performed Wanda's autopsy had been able to identify it when he examined her brain tissue under a microscope.

Gail asked if it could be related to whatever had been wrong with Moe's mother, who had died at a young age with similar symptoms; the doctor said possibly. He also said that if that's what it was, maybe half of her children could inherit the disease, which was the theory that Robert Feldman's 1963 paper had suggested. It was a crucial detail that Gail, to her everlasting regret, would not fully grasp for another thirty years. She thought the possibility of the disease being hereditary was much more remote, the way any other ailment could be—something that a few of the kids might get when they were old, the way some people might worry about heart disease if a parent suffers a heart attack, or about cancer if it seems to run in the family. There was no medicine the doctor could offer.

Back in Tioga, the DeMoe house remained a hub of activity—it could hardly be otherwise with such a large family. But as Moe's forgetfulness increased, his personality changed, too; always strict, he became angrier, more irrational. Moe's temperament would forever be a point of dispute between Gail and her children. They saw him as a mean man, but she recalled a different Moe, the one who adored children and had been proud of his large brood. He had never been a soft touch, but it was the disease, she insisted, that changed him into a moody, violent man. The kids weren't buying it—not even Karla, his favorite child. His personality changes had occurred so gradually that it didn't seem as though they happened because he was sick; it seemed, to their teenage eyes, like he was behaving that way deliberately, because he didn't like them. He'd been acting that way since their preteens; in Jamie's case, because he was the youngest by so many years, it was virtually his entire life. They began avoiding the house, and so did many of their friends.

At work, Hank Lautenschlager was slow to notice any changes. "I was so close to him, I didn't realize what was going on till it was too late," he admitted ruefully. "I just couldn't believe that he was like that."

The first time Hank realized that something might be amiss was when Moe got angry with the shop foreman over some perceived problem. "Moe came off the truck with a hammer and chased him around the truck," Hank recalled.

Although the foreman forgave him and they went out for a drink that same afternoon, the truce didn't last long. In the oil fields, if you couldn't do the job, there was only so long anyone would carry you—even if you were as well liked as Moe. Drill stem testing was the last upper-level job he'd hold; he returned to driving a truck. That lasted until he started struggling to find locations and ran over a mailbox. Then the company demoted him to washing the trucks he used to drive.

"He'd stand there and just wash the floor down for hours," said Hank, who had an unwilling front-row seat to Moe's decline. One night, Hank looked out his window to see Moe poking around his own house, looking for a lost set of keys. He was naked.

Eventually, in 1974, Moe was let go from the job. It was a devastating blow, one that cut to the core of his self-image as a hardworking man who supported his family, who could be trusted in the fields. And while he drew some disability pay, the financial impact was significant, too. The family went on food stamps. Karla, ever conscious of appearances, was so ashamed that she refused to go to the store for her mother, unwilling to admit to government assistance in front of friends who were bagging groceries.

To help support the family, Gail took a job as a nurses' aide at the Tioga hospital, and Moe stayed home to take care of three-year-old Jamie. The older children weren't interested in watching their youngest brother—there was too much fun to be had out on the town. For most of his childhood, Jamie would be an afterthought to his older siblings. When he was a baby, nobody would share a room with him, so his crib was parked in the hallway outside Lori's room.

On nights when Gail had to work a late shift, Moe sometimes forgot where she'd been and worried she was cheating on him. Alcohol made things worse. In Gail's memory, he didn't take out his anger on the kids—just on her.

"I was the one he looked to for answers," she said. "I guess you pick on the ones you love the most."

But most of the children told a different story. Except for Karla, they all remembered being on the receiving end of his abuse. As much as they loved their mother, they could not tolerate their dad. And the worse he got, the more it tore the family apart. Brian escaped when he graduated high school and moved to Wahpeton, four hundred miles away, to attend the North Dakota State College of Science. Unbeknownst to his siblings, he stayed in touch with Moe—possibly to ask him for spending money without Gail finding out. Feeling increasingly trapped, the other children also began looking for a way out.

In the early Sunday morning hours of a summer heat wave, Lori woke to the sound of her father, "drunker than hell," yelling at her mother. When Lori told him to quiet down so she could sleep, he came after her, and she locked herself in the bathroom. She would recount it later in a letter:

> *He started after Mom again + she sprayed him in the face with a can of tear gas that the boys had bought. We all got the gas but Dad got it the worst. He ran outside to get away from it. He didn't have any clothes on. When he came back into the house he was really, really pissed. He started to hit + chase Mom and I grabbed him to hold him back cuz I was scared and didn't know*

what he would do if he would have caught Mom. He threw me up against the stove but I grabbed him again until Mom could get out of the house.

With none of her brothers home to help her handle Moe, Lori called the police. When the officers arrived, Gail retreated to the neighbors' house while they calmed him down and got him back to bed.

Mom really doesn't know what to do now. She might put him in some kind of a hospital. I think it should have been done awhile ago. It's tough on everybody.

Eventually, Lori left. She found her ticket out while cheering at a high-school wrestling match, where she fell in love with one of the spectators, Steve McIntyre, an out-of-towner from the Pacific Northwest who was visiting family. From the moment they went on their first date, Steve knew he wanted to spend the rest of his life with her.

After she met Steve, Lori had one foot out the door. While he crisscrossed the western United States working as a track maintenance man for Union Pacific Railroad, she finished high school, started trade school, and then set out to join him.

She and Steve were both nomads at heart, and she had never been the kind of person to worry about what others thought of her. In trade school, she chose an auto parts program, unusual for a girl—particularly in 1977. In one letter to Steve, she wrote about the new outfit she was sewing for herself—cursing her way through her mistakes—before she abruptly switched topics to the automobile power train system she was studying. Tioga would gradually recede in Lori's rearview mirror as she and Steve embarked on what she was sure would be a life of adventure. She left her siblings back home to figure out how to coexist with Moe as best they could.

The shouts behind the closed bedroom door continued; Moe seemed to be pushing Gail around. But in her room down the hall, Karla stayed in bed. She didn't see what her father was doing to her mother, and she didn't intervene; neither did the boys, holed up in their basement bedrooms. All she knew was if she was headed home at the end of the day and her father's car was in front of a bar uptown, it was going to be a bad night.

Karla isolated herself from her siblings, spending as much time as she could with her friends. As an adult, she would regret the lack of support they gave one another during the worst period of their young lives, but she believed they were simply shutting down to survival mode, because they had once been so close.

"You're just trying to make it through it all yourself. That's all you think about: This is awful for me," she said.

But the family bond that Gail had created was stronger than any of her children understood at the time. Although they were beginning to scatter, trying to forget the rough parts of their upbringing, they never fully broke free of their roots. And when life soured for them at times, they sought comfort in one another. It was a quality each of them would strive to instill in their own children.

Meanwhile, after Lori's departure, Dean and Doug were still raising hell around town, and Brian was still in Wahpeton. Karla attended college for a year, then had returned to Tioga to work an administrative job for the town when the escalating situation with her father finally detonated.

Gail asked Brian to move back home, thinking he could help handle Moe; it didn't work. In fact, with one more person to confront him, Moe became, if anything, more irrational. Only the police could settle him.

In August 1978, Karla was at work when the police department's phone rang over to her desk, as it sometimes did when nobody else was there. It was someone from the sheriff's office calling to say they were sending a car to pick up Galen DeMoe.

Karla's heart sank, but she finished the conversation with the dispatcher and quietly hung up, never revealing that it was her father they were discussing. The shame bubbled in her chest. When would it ever stop?

Gail had run to Hank's house to hide, and from there had called the police, hoping they'd be able to tame her husband once again. But instead of calming down, this time Moe's rage increased. He fought with the officers until the police finally managed to shove him into the patrol car and drive off down the street. Dean, then sixteen, huddled in front of the window, convinced his father would break free and come after him. He didn't exhale until the cruiser drove away.

The next day, Gail called Karla, who could hear the fear and pain in her mother's voice.

"They're taking Dad away for good, for my safety," Gail said.

Moe's sons packed his suitcase for him. A judge committed him to the North Dakota State Hospital in Jamestown, 250 miles away, a facility for the mentally ill and criminally insane. Galen DeMoe never came home again.

Three

FAMILY N

IN THE YEARS since Auguste Deter's death, thanks to Emil Kraepelin's decision to define Alzheimer's as a malady striking people under the age of sixty-five, science continued to mistakenly treat it as a rare affliction, separate from the common senility of older people. That error cloaked how widespread Alzheimer's really was.

Things began to change in 1948, when a London scientist named R. D. Newton reviewed the previous forty years' worth of literature written about the disease and concluded that there were no grounds for distinguishing between Alzheimer's and senility. Newton also found previous reports that Alzheimer's appeared in families in German, British, and American papers. When he grouped those isolated reports together, he became convinced that a hereditary factor contributed to the disease, whether the patient was old or young. Two more studies in the next dozen years would work to solidify Newton's idea that a genetic link was at play, including the one written by Harvard's Robert Feldman in 1963, the year before Moe's mother, Wanda, died.

But the turning point came in 1968, when a group of British scientists—Gary Blessed, Bernard Tomlinson, and Martin Roth—showed that most people over sixty-five who had dementia also had the same plaques and tangles that had overrun Auguste Deter's brain, firmly establishing that senile dementia and Alzheimer's were the same condition. Their research debunked the prevailing theory that most senility was caused by atherosclerosis, or what is sometimes called "hardening of the arteries," when fat builds up inside artery walls, causing them to narrow and stiffen. The only thing rare about Auguste's condition had been that it struck twenty years early.

Once Alzheimer's researchers understood that they were dealing with a much larger patient population than they had realized, they began making tentative steps toward discovering the disease's causes, including possible genetic factors. But it was hard-won knowledge, collected from small pockets around the world when doctors stumbled across cases so strange, and sometimes so poignant, they demanded further investigation.

On the southern tip of Italy—basically the toe of that famous boot—is the region known as Calabria. Its unstable, mountainous terrain is prone to violent expressions of nature, notably volcanoes, earthquakes, and tidal waves, which have killed tens of thousands of its inhabitants over the centuries.

It was into this beautiful, sometimes brutal land that French neuropathologist Jean-François Foncin landed in May 1973, hoping to unravel the strange family history of his patient, a forty-four-year-old woman with Alzheimer's disease. What he did not know was that his journey was going to dramatically expand science's understanding of the genetic components of the disease, and the way it is inherited.

A year earlier, the woman—an Italian immigrant who was then living in France with her husband, a tile layer—had gone to a doctor when she had difficulty caring for her newborn, who was her ninth child. Her symptoms pointed to some kind of brain disorder, and she was referred to a neurologist who committed her to a psychiatric institution. She was increasingly disoriented, and doctors struggled to evaluate her memory and speech since she was illiterate and spoke only in her native Calabrian dialect. Suspecting a possible brain tumor, they transferred her to a Paris hospital; in those days before CT scans allowed for a less invasive diagnosis, the surgeons instead drilled a burr hole in the right front section of her skull—an ancient procedure known as trephination—and drained a sample of cerebrospinal fluid from her brain's ventricle, searching for possible tumor cells. They found none.

Doctors then performed a cortical biopsy, which means they removed a small sample of brain tissue. Brain biopsies are not undertaken lightly; complications can include seizures, brain infections, even death. But when symptoms are baffling, particularly in young patients, they are sometimes used to diagnose (or rule out) potentially treatable illnesses, such as brain inflammation.

Unfortunately for the woman from Calabria, the sample revealed something much more insidious: plaques and tangles.

"There's no tumor here," Foncin told his colleagues. "It's Alzheimer's disease."

Foncin recalled reading the early theories—from Newton and others—that there might be a form of early-onset Alzheimer's that ran in families in an autosomal dominant pattern. So he asked the woman's husband about her family history. The husband promptly told him her sister and father had died of the same disease.

Suspecting he was onto something, Foncin sent a request to Italy for her father's medical history. He received three sets of records in response: those of the father and two other people with the same surname, all of whom had died with dementia in their fifties.

The discovery also piqued the interest of Foncin's mentor, Dr. J. E. Gruner, a prominent French neuropathologist and amateur genealogist. Gruner arranged for private funding to further investigate this possible genetic link to Alzheimer's disease—which held the potential to explain both how it happens and predict when it might strike next. Together with Gruner, Foncin trekked to Calabria to see what he could find.

The patient's relatives, who would come to be known as "Family N" in medical lore to preserve their privacy, were cooperative. By now, the patient had returned to Italy, where she was "profoundly demented and cared for like a baby by her old mother," wrote Foncin. The mother was also caring for other siblings and cousins with the same symptoms. Municipal records and archives at the local psychiatric hospital yielded a wealth of information on the family dating back to 1880.

In village churches, provincial Italian priests traditionally kept a journal of the preceding year's events, annual write-ups summarizing their parishioners' triumphs and trials: *"Giuseppe has started wandering, quite confused; couldn't find his way back home, just like his father."*

The woman Foncin saw in Paris had lost her father, grandfather, and great-grandmother, all of them dying in similar ways: in their fifties, with dementia. The diagnosis would read "syphilitic general paresis of the insane," but the family simply called it "the disease."

To further pursue the autosomal dominant inheritance of Alzheimer's in Family N, Foncin applied in 1977 for much-needed funding through a grant from the French Institute of Health and Medical Research, which was running a program on brain aging. Suspecting that his case might not be exactly what the institute had in mind for its program, Foncin didn't offer many details about his project. He won the five-year grant on the stipulation that he present his findings halfway through to a committee set up by the institute.

Using punch cards and a mainframe computer, Foncin and a colleague began to quantify the genetics of Family N. When the time came for his presentation, he was so confident that the family trees he was constructing were illustrating something valuable that he decided to tell the committee exactly what he was up to. He presented extended genealogies traced through the computer and added his opinion that this was the way to solve the Alzheimer's riddle.

Disappointed, the committee members did not agree. They had been hoping he was searching for a way to delay the Hayflick phenomenon, which is the number of times a human cell population will divide before it finally dies. The committee cut off Foncin's grant.

Demoralized, Foncin put aside Family N for months. But then Gruner learned that one of his former students had established a neurology unit in the hospital that had served as the birthplace of Foncin's original patient. Foncin's passion for the topic was rekindled; he flew to Calabria and reached out to one of the doctors in the unit, who agreed to help him with his research.

They discovered more relatives afflicted with the disease, with branches extending into other countries, including the United States and France. And they also realized that they weren't the first doctors to take an interest in the family. It turned out that when Robert Feldman wrote his 1963 paper suggesting that genetic inheritance played a role in Alzheimer's disease, he was basing his theory on the research he had done with members from a branch of Family N.

Sometimes, doctors who work with extended families become very close to their subjects; such was reportedly the case with Feldman, who died in 2003. He watched helplessly as members of Family N, even the brightest and most high-achieving of the bunch, succumbed to Alzheimer's at unusually young ages—in at least two cases, they were thirty-one and thirty-two years old when their symptoms began.

Feldman did the best he could to analyze the family's heredity with mid-1960s technology, but it wasn't enough. When Foncin's group contacted him, however, the answers began unfolding. In 1988, a quarter of a century after Feldman first reported on the family, he, Foncin, and several collaborators were able to trace their subjects' roots all the way back to a single common ancestor: a woman named Vittoria, born in Calabria in 1715, who had died at the age of forty-three. They had sixty cases of Alzheimer's disease spanning ten generations of Family N, and they had identified four thousand descendants across several countries. Their 1985 paper appeared to offer conclusive evidence of a form of Alzheimer's disease that was repeating itself genetically across the generations, and the disease was the same regardless of the country where the victim had grown up. It was a treasure trove of genetic information.

In the United States—in fact, just a few miles from where Feldman worked at Boston University—a wry young British neurologist and molecular geneticist named Peter St. George-Hyslop was just getting started in his career. As he was finishing his medical training in 1985, he took a position at Massachusetts General Hospital in a lab specializing in neurodegenerative disorders. Since Hyslop was interested in the DNA analysis of Alzheimer's, it seemed a good place to start.

In the Mass. General lab, Hyslop found write-ups of a Canadian family with Alzheimer's whose DNA samples had already been submitted to the National Institutes of Health (NIH). While flipping through a book of the samples, Hyslop found a reference to Alzheimer's disease on slides labeled with an Italian surname. But when he asked about them, nobody knew where they had originated; they had become the scientist's equivalent of the mystery contents in the back of Gail DeMoe's refrigerator. Hyslop left them there.

During the next few months, Hyslop struggled to find additional families with autosomal dominant Alzheimer's so he could create a large enough sample size for meaningful research.

"I was aware that there was this other Italian family from Foncin, but I had never met [Foncin], had no idea how to get hold of him," Hyslop said.

Then, his luck turned. Rick Myers, a researcher who gave genetic counseling to potential victims of Huntington's disease, wandered into the lab and asked whether Hyslop would accompany him to offer a neurological opinion on a case.

As the two men drove to the patient's house, Hyslop explained that he was looking for families with Alzheimer's disease. He mentioned the mysterious orphaned samples he'd found.

To his complete astonishment, Myers said he himself had collected some of the samples and knew all about their origins. They were from Family N, one branch of which had immigrated to Massachusetts in the 1950s.

It was a eureka moment: Suddenly, those biological leftovers were extremely useful. Instead of struggling to find enough people to cobble together a respectable sample size, Hyslop now had a large enough family tree to roll up his sleeves and pinpoint whatever biological properties set them apart. Had they never contributed their DNA, he would still have been seeking genetic carriers, one by one, hoping to isolate the gene that was afflicting them.

Hyslop dove enthusiastically into his study of the Family N tree. A few months later, he was invited to give a talk at a meeting in France; there, he finally met Foncin, and he explained that they were working on two different corners of the same enormous genetic jigsaw puzzle. It was the beginning of a long and productive collaboration among the American, French, and Italian research teams.

At last, Jean-François Foncin had found a kindred spirit: someone who recognized the scientific value of a family who had been so tragically stricken. His collaboration with Hyslop and their work on expanding the Family N tree became well-known, enough to make Foncin a minor celebrity in the world of French genetic and Alzheimer's research. The same committee members who had cut his grant nine years earlier looked at his project with a sense of regret, and said, "If we only had known."

Finally, researchers were on the right track to identifying a genetic component to Alzheimer's—and perhaps, in doing so, better understanding the mechanics and life cycle of the disease, well before it was visible to the eye, which could tell them what mechanisms to target, putting them on the path toward a cure.

Four

ONE IN A MILLION

FOR SEVEN DECADES, research funding for Alzheimer's was extremely scarce, since the disease was considered so rare. The 1968 discovery by British scientists Blessed, Tomlinson, and Roth that the common senility of older people was actually Alzheimer's helped the field begin to realize, for the first time, the magnitude of the situation. But better research funding didn't materialize immediately after that discovery. In fact, it would be 1976—eight years after their paper was published, and three years after Foncin landed in Italy to research the genealogy of his unusual patient—before a doctor in San Diego issued a call to arms.

Dr. Robert Katzman, a neurologist and medical activist from the University of San Diego, wrote an editorial in 1976 for the *Archives of Neurology* that called Alzheimer's "a major killer," identifying it as the fourth-leading cause of death in the United States after heart disease, cancer, and stroke. (It later fell to sixth when chronic lower respiratory diseases and accidents were factored in.) The following year, he organized a conference on Alzheimer's that won support from the National Institutes of Health. He believed the only way to effectively fight the disease was to begin building an infrastructure that would allow researchers to attack Alzheimer's from multiple angles.

In April 1980, Katzman helped found the Alzheimer's Association to support federal research related to the disease. His efforts took time, but they paid off: From 1980 to 1996, federal funding for Alzheimer's research increased from $5 million to $300 million. Though that amount still fell far short of the money spent on other diseases, the momentum built in that time frame resulted in the most detailed knowledge yet gathered about Alzheimer's, as well as the first attempts to develop drugs that might treat it.

Now that Family N had helped establish a genetic link to Alzheimer's, the next logical step to deconstructing the disease was to locate which specific genes were affected by mutations, and then duplicate the mutations themselves by cloning them for closer study. In the mid-1980s, that was no easy task. Recognizing that it was a crucial battle, teams of researchers raced one another to become the first to clone an Alzheimer's gene. But it would take several remarkable coincidences of plain good luck, and the faith of a newly arrived Russian immigrant, before that milestone was achieved.

Dmitry Goldgaber was up for the challenge.

Born in Latvia in 1947, Goldgaber was the son of a Jewish father and a Russian mother. As a child, he won prizes in math and physics competitions. During high school, he read *Brighter Than a Thousand Suns*, Robert Jungk's book about the builders of the atom bomb. From that point forward, he was determined to become a molecular scientist.

Goldgaber graduated from high school and immediately made plans to apply to Moscow State University, the most prestigious institution in the Soviet Union. His parents warned him that an applicant with a Jewish last name would never be accepted in the pervasively anti-Semitic climate of Soviet Russia. But young Dmitry was confident that math was the great equalizer. He would pass the entrance exams with flying colors, and Moscow State would welcome him with open arms. They had to.

After the first round of tests, he appeared to be right. Out of thousands of applicants, he was one of the few who passed, and he earned one of the higher grades. But when he took the oral portion of the entrance exam, he was shocked when the examiners failed him. Undaunted, he appealed; and the department chairman handled the petition personally, as if to leave no doubt about the finality of the decision. He asked the young student questions from every discipline in math, inside and outside the scope of his program. For several rounds, Goldgaber answered flawlessly.

Finally, the chairman found a question Dmitry could not answer.

"I don't know," he admitted.

"You see? You fail!" the chairman answered. His parents had been right; his last name would keep him out of Moscow. It didn't matter that his mother was Russian; in the Soviet society, any Jewish heritage, no matter how remote, was grounds for discrimination.

Though bitterly disillusioned, Dmitry picked up the pieces and started over. His mother reminded him that he'd promised her if he failed, he would try again somewhere else. His exam scores were high enough to get him into Latvia State University and, later, Leningrad Polytechnik Institute.

One summer during his stay in Leningrad, he worked on a construction crew as part of a Soviet program that dispatched young people to remote areas to work on state-sanctioned building projects. Goldgaber was shipped to western Siberia to help construct buildings for crews working in natural gas exploration, and the trip to Siberia would change his life. An hour into the flight, he looked down and saw a series of perfect squares far below on the ground. When they landed, he asked the pilot what they were.

"Prison camps," the pilot explained. They were empty—built just in case the government decided it needed them.

"I was absolutely horrified," Goldgaber said. "At any moment, someone up there could push the button, and thousands could go work in those camps. I said, 'I have to get out. I can't live in a country like this.'"

He graduated in 1978, but he was married and had a son by the time he finally got his wish. Then a junior researcher, he applied for government permission to emigrate.

As it happened, the anti-Semitism that had kept him out of Moscow State University a decade earlier would help him now. In 1974, the United

States had passed an amendment that linked the most favorable US tariff rates to the rights of Soviet Jews to emigrate freely. At the end of 1979, Goldgaber was among thousands of Jews who were permitted to exit the country. (He was exceedingly lucky. Just three weeks later, the Soviets invaded Afghanistan, souring the US trade negotiations and prompting the Soviets to once again slam their doors shut.)

After his narrow escape, Goldgaber flew with his family to Vienna, where he set about writing letters to three scientists who he'd been told might help him find a job.

"Most likely, I will start with being an attendant in a gas station," Goldgaber recalled thinking. He didn't know how to drive, so he knew he couldn't be a cabbie. "But gas station, I thought, 'Yeah, I can manage.' "

One of the letters was addressed to Carleton Gajdusek from the National Institutes of Health. Gajdusek had built a career out of studying genetically isolated populations for clues about rare illnesses. Three years earlier, Gajdusek—pronounced GUY-dah-shek—had won the Nobel Prize for identifying kuru, also known as the "laughing sickness," an exotic disease affecting a full 10 percent of a tribe in Papua New Guinea.

Its victims shook and broke into fits of laughter and madness before dying; autopsies revealed their brains were shot through with gaping holes, like sponges. Gajdusek linked the disease, which was previously thought to be caused by heredity or dietary deficiencies, to the tribe's ancient funeral rite dictating that women and children show their respect for the dead by eating the corpse's brain. That custom, Gajdusek concluded, planted a time bomb: The slow-acting infection that it spread would explode several years after the funeral.

From Austria, Goldgaber traveled to Rome, where Gajdusek's polite response was already waiting for him. Goldgaber's limited English didn't allow him to fully understand it, so he took it to the Hebrew Immigrant Aid Society, which helped Jewish refugees. There, a translator explained that Gajdusek was offering him a job in his lab at the NIH.

Goldgaber was ecstatic; he could hardly believe his luck. Not only would he not have to pump gas, he would be practicing science with a Nobel laureate in one of the United States' most revered institutions. All he needed now was verification through the refugee service that the offer was legitimate.

Unfortunately, Gajdusek—who spent much of his time traveling around the world in pursuit of his science—was a hard man to reach. In that era before cell phones, pagers, or social networking, trying to reach a traveling scientist who specialized in isolated populations was nearly impossible. As the days went by, the refugee service began talking about sending Goldgaber someplace else.

Desperate, Goldgaber decided to try calling Gajdusek's lab himself. He went to the central telephone exchange in Rome to place the call, ashamed that he had to make it collect.

A woman on the other end answered, but he couldn't understand her machine gun–rapid English.

Goldgaber tried again: "My . . . name . . . is . . . Dmitry Goldgaber. Do . . . you . . . know . . . my . . . name?"

"Hold on," the woman answered.

Goldgaber stared at the receiver in his hand. "I didn't know what 'hold on' means," he said. "So there was silence, and I said to myself, 'What should I hold?' "

The interminable quiet continued. But Goldgaber told himself if he hung up, it was over—no NIH, no job in science, back to pumping gas. Finally, the receiver came alive again. And then, a miracle: the male voice on the other end spoke in Russian.

"Wait. Don't go anywhere else!" the lab employee said. "I will come to Washington and pick you up."

In March 1980, Goldgaber came to the United States. He was thirty-three years old, and like many immigrants, he refers to his arrival date as "my second birthday."

Carleton Gajdusek wondered if Alzheimer's—like his earlier discovery, kuru—might be a prion disease, which is contagious. Prions are protein particles that are naturally produced in the body and are thought, among other functions, to allow brain cells to communicate with one another. In their abnormal form, they fold over and infect other prions, causing them also to fold, which eventually kills the brain cells and leaves a series of holes in the brain. Mad cow disease, scrapies, and some forms of Creutzfeldt-Jakob disease are all examples of prion diseases, which cause a breakdown in memory, personality, behavior, and physical and intellectual function. Victims of prion diseases, like Alzheimer's patients, get amyloid in their brains.

In 1980, Gajdusek tested the theory by trying to infect chimpanzees with Alzheimer's brain extracts, but the experiment failed. Scientists began looking for other causes.

In 1984, a pathologist named George Glenner from the University of California at San Diego identified the exact makeup of the protein that forms the amyloid plaques found in the brains of Alzheimer's patients.

He found them to be much smaller than the amyloid of prion diseases, but they were identical to plaques found in the brains of people with Down syndrome—one of whom was Glenner's daughter. He decided to pursue this coincidence further.

Both Alzheimer's and Down syndrome plaques are made up of amyloid beta, a fragment of the amyloid precursor protein, or APP, manufactured by chromosome 21, an extra copy of which is what causes Down syndrome.

Virtually all people with Down syndrome develop amyloid beta plaques in their brains beginning in their thirties, and while not all develop Alzheimer's, studies estimate as many as 75 percent do.

Based on his findings, Glenner theorized that at least one of the defective genes behind Alzheimer's disease, causing amyloid beta to run haywire through the brain, would be found on chromosome 21. To prove his hypothesis was true, someone would need to find that gene.

Genes instruct cells about how to build a particular protein. A scientist looking at a specific gene can identify the proteins it dictates with relative ease. And now that the human genome has been mapped, the reverse is also true: It's simply a matter of using a computer to search through the human genome database to find the originating gene from a known piece of protein.

But back in the early 1980s, there was no database to search; the quest would begin by painstakingly cloning genes one by one, then checking to see if any of them matched the known piece of amyloid beta protein. The numbers were daunting. There are roughly twenty thousand to twenty-five thousand genes that code proteins in the human body, and each one can code multiple proteins. So finding this particular amyloid gene was like searching for a four-leaf clover, blade by blade, in a thousand-acre meadow. Having a hint that the mystery gene was on chromosome 21—because of its link to Down syndrome—could offer a significant shortcut.

Within a few years, Goldgaber started working on neurodegenerative diseases of the brain. He noticed the scholarly interest his new boss, Gajdusek, had taken in Alzheimer's disease.

Most of the people working in Gajdusek's lab were visiting fellows from other countries; very few actually worked for the NIH, though Goldgaber was an exception. Gajdusek liked to excite them with scientific conundrums. If someone in his lab felt passionate about a problem and committed to pursuing a solution, Gajdusek felt the job was halfway done.

Such was the case in 1985, when Goldgaber approached him about the idea of trying to clone the mysterious amyloid gene. Goldgaber had learned gene cloning as an unpaid apprentice for a well-known Soviet scientist and expert in molecular biology who was working next door at the National Cancer Institute.

Goldgaber's proposal was hardly original, but he didn't know it. In fact, more than twenty research teams around the world were trying to do the same thing; the race was on. Had he known, Goldgaber never would have bothered. Noncompetitive by nature, he preferred instead to go against the grain: "If so many people are working on the problem, it will be solved. I work in science not for competition, not because I want to be first in the finish line, but because I am curious about this problem."

He didn't know anything about the disease that had prompted the flurry of interest; he couldn't even spell "Alzheimer" without looking it up. It didn't matter. Gajdusek agreed to let him try, although he later admitted that he assumed his eager young protégé would fail.

So Goldgaber purchased commercially made libraries—pieces of brain DNA that arrive pre-sliced and laid out in petri dishes—and made some of his own, then set to work creating a probe, which is a single strand of DNA that has been made radioactive. When it matches another DNA sample, it sets off a glowing signal that is captured on X-ray film.

Goldgaber's then wife, who had emigrated with him, worked for a computer firm that serviced the NIH. She ran a computer sequence comparing amyloid to a database of known proteins and protein sequences. When they found a strand that was unique to amyloid beta, they fashioned it into their probe.

For several months, Goldgaber's efforts often went comically and absurdly wrong, as though fate was pranking him. The probe he designed used a small molecule called deoxyinosine, which was only made by two companies in the United States at the time. Unfortunately, Goldgaber happened to pick the manufacturer that, soon after he contracted with them, suffered an explosion in the room adjacent to where the deoxyinosine was produced, delaying his probe for an interminably long period.

To kill time while he waited, he enrolled in a class at the University of North Carolina to master sequencing, a technique in analyzing DNA and proteins. On his drive south, his car broke down.

Still unaware of the global competition, Goldgaber nonetheless worked feverishly: At one point in 1986, he emerged from his lab during daylight and was surprised to find that the seasons had changed and spring had arrived.

Screening DNA libraries without a computer or a map of the human genome is mind-numbingly tedious work. Goldgaber would lay the DNA samples in a large petri dish, then introduce the probe, then check the X-rays to see if there was any signal indicating a match. Sometimes he would see weak signals, raising his hopes, but they sputtered out each time. The

X-rays looked like negatives of the sky at night, spattered with stars: The background was clear, and tiny dots covered it, each representing a piece of DNA that didn't contain the gene responsible for amyloid beta.

By June 1986, having tried and failed with both commercially made and home-grown libraries enough times to discourage even the most dogged optimist, Goldgaber was down to one final library. It was a weekend, and Goldgaber had nothing else to do ("Scientists are usually bad parents," he said). Besides, he hated to miss any opportunity, however remote.

"All my life, I've had this approach," he said. "If there is one chance in a million, take it. Try it."

With lukewarm enthusiasm, he screened his last library.

And then, he hit the jackpot: There was a huge signal on the X-ray film.

In the constellation of DNA fragments, "this positive signal was, like, the largest planet of them all," said Goldgaber. "Little tiny dots all over, and then bang! A big black spot."

Not sure if he should believe his eyes, Goldgaber went back to his petri dish and cloned more examples of the microscopic fragment that had given him such a strong signal. Eventually he had a dish with nothing but the same fragment in it.

When he took X-rays this time, he found the same large spots, repeated over and over, wherever the amyloid gene lay. Goldgaber translated the DNA into its protein sequence, and there it was: the sequence of George Glenner's amyloid beta protein. Just as Glenner had predicted, it was on chromosome 21, the same extra chromosome found in people who have Down syndrome. That was why both Down's and Alzheimer's patients had so much amyloid beta in their brains. And why people with Down syndrome who lived past the age of forty so often developed Alzheimer's.

When the amyloid precursor protein (APP) gene is mutated in Alzheimer's patients—and there are about fifty different types of known mutations that occur there—it causes people to get the disease at a very young age, usually in their forties.

But APP is not alone. At least two other genes also mutate and cause

Alzheimer's, and science continues to search for others. What they all have in common is this: They result in dementia, whether you get it in your thirties, your forties, or your eighties.

"That's it!" said the man who had missed the coming of spring while he toiled in the lab. "I am not working today anymore."

He drove over to Gajdusek's house, where they shared the most expensive cognac the local liquor store carried.

Gajdusek urged Goldgaber to present his findings at a session of the Society for Neuroscience meeting in Washington, DC. Goldgaber showed up and, after some negotiation, was allowed to preempt the program and present his data as an unscheduled guest. In heavily accented English, he briefly explained his discovery to the packed audience, referring to transparencies as a visual aid. Word had spread that an unknown had found an amyloid gene; people sat in the walkways while still more squeezed through the door.

While he spoke, the crowd was completely hushed; the silence confused him, because he had expected questions. As with Alois Alzheimer more than eighty years before, it seemed as though he had flopped.

"If there are no questions, I thank you for your attention," he said, and prepared to leave the podium.

A man in the third row stood up, scratched his head, and said loudly: "What do you mean, there are no questions? Of *course* we have questions! Lots of questions!"

Suddenly, the room was electrified. Unlike the underappreciative audiences of Alzheimer and Jean-François Foncin, these were people who finally had an idea of what they were up against. They weren't fighting a statistically rare disease, and they weren't fighting the inevitable consequences of old age. United by the rallying cry of Robert Katzman, they were fighting a public health crisis, propelled by the discovery that generations of families were genetically linked to the disease. Goldgaber had just advanced them an important step forward.

Goldgaber answered questions as best he could, then stepped down and walked into the hall. Everywhere he went, people congratulated him, patted his shoulder, asked him for copies of the gene. He was feted, offered jobs in start-up companies dedicated to Alzheimer's research. Goldgaber politely declined; academia was his home.

But he had every reason to celebrate: He had seized the education that Soviet society would have denied him, escaped the oppression in his homeland, and found a job working in his field when so many other immigrants were forced to start from scratch.

"The chance I got when Carleton hired me: one in a million," he said. "I was not known. Just a junior researcher, an immigrant from Russia with one publication, no paper in *Science*, nothing. . . . I understood very clearly that for me, it was incredible luck."

A quarter of a century later, he still gets emotional at the thought.

Five

YOU ARE MY SUNSHINE

AFTER A JUDGE committed Moe to the state mental hospital in Jamestown, Gail spent the next ten years in an extended state of limbo. It was an uneasy peace for a woman who could no longer tolerate her husband's abuse but who had spent the past twenty-three years being a wife and mother.

"I remember asking my mother: 'I'm not a widow. I'm married, but he's not here. I don't know what to call myself,' " she said.

"Call yourself a semi," her mother answered. "You're semi-married."

Though the family was well liked, their decision to commit Moe to an institution hadn't sat well with some folks in town, court ruling or not. Even a few family members disapproved. Small-town judgment can be hard to handle. But the people behind those whispers didn't have to live Gail's life.

Only she fully knew the terror of Moe's late-night rampages, of being shoved around while her children listened to muffled shouts through bedroom doors. She knew the desperation of running to her neighbors for help while he fumbled in the bushes, naked, searching for nonexistent keys; of covering up the previous night's scene with pranks and self-deprecating jokes. She knew the injustice of explaining to the man she loved that she was not cheating on him. She knew the helplessness of seeing him waste away, hundreds of miles from home, while he begged to see the family who had no way to take care of him, who didn't remember him the way she did—the affectionate, rough-hewn man who had won her love as a teenager.

Night after night, Gail pored over the paperwork associated with Moe's illness, while Tioga grumbled that she should have been taking care of him herself. The emotional strain, coupled with the growing financial burden, eroded her once unsinkable optimism. Shortly after Moe left for Jamestown, she suffered the first of a series of nervous breakdowns and was hospitalized in Minot for about a week. Depression always seemed to hit her hardest in the spring; she wouldn't be able to get out of bed, and Karla could hear the despair in her mother's voice.

"She would be so mad at herself because she just couldn't pull herself out of it," Karla recalled.

Moe spent four months in Jamestown, which Gail euphemistically referred to as "a learning experience," because Jamestown housed everybody, regardless of where they fell on the mental-health spectrum. No matter what your problem was, or if—as with Moe—nobody was sure what your problem was, Jamestown would take you. It was an imposing redbrick building, drab inside and out. When patients first arrived, they were all thrown together on the first floor before being redistributed.

Gail was afraid to visit Moe at first, in case he blamed her for his situation, even though the entire matter had been decided by a judge. But when she did go to see him, she found him passive, almost plaintive. He apologized for hurting her and begged her to take him back.

"I'll be good, I'll be good! I won't get after you anymore," he pleaded. It wasn't clear whether he actually remembered those hellish nights or was simply apologizing for what he'd been told he'd done.

But even if the judge had permitted Moe to return home, the family wouldn't have been able to handle him. Gail did have some good news for him, though: Thanks to Moe's military service, the veterans hospital in St.

Cloud, Minnesota, had agreed to take him. It was eight hours away from Tioga, but it was still better than the cheerless pallor of Jamestown.

St. Cloud would become Moe's home for the next seven years, from December 5, 1978, until January 8, 1986, when the staff determined that his condition warranted a move to a nursing home. For the first year he was there, he asked constantly to return to Tioga, although his family visited him as often as they could.

"I want to go home and be with my kids," he pleaded. A nurse reported that he was "sitting in the dayroom sobbing as though his heart would break."

"I can't help it. I want to talk to my wife," he said.

But none of his children would know any of those details until decades later, when Karla requested his medical records. By August 1980, Moe could not remember the names of all his offspring. A year later, he had stopped asking about them. By April 1985, he could not consistently respond to his own name. He sat strapped into a chair designed for geriatric patients, alternately laughing and crying.

Back in Tioga, unaware of their father's heartbreak, the boys did their best to heal, as though they were finally able to exhale. Gail, her emotional reserves depleted, focused on keeping what was left of her family as intact as possible—even expanding to include her children's friends. But that didn't leave much energy for her to focus on discipline. That had always been Moe's department, anyway; now that he was gone, the boys tested their limits.

In the detached garage in the backyard was Moe's car, a newer model that nobody had ever, ever been allowed to touch. "I think the day Dad left, that car went down the road," Karla said. Not long afterward, Dean flipped it over into a water-filled ditch. A picture of the wreck made the newspaper. Soon after, Brian rolled his own car on Main Street, smashing into some storefronts.

Part of their rebellion was fueled by their need to show the rest of Tioga, as visibly as possible, that they were stronger than what they had endured. No matter how embarrassing or degrading Moe's behavior had been, it had not broken them. They were still the most entertaining family in a town where a certain amount of lawlessness was tolerated, even admired.

Rather than worrying Gail even further, the boys' antics provided welcome comic relief. They kept her young, reminding her what it was like to laugh. The indignities of Moe's illness fell away then, and she was, once again, the madcap Gail who was always ready for anything. Karla admired her for that.

Brian's sense of humor and love of practical jokes deflected much of the anger and pain he experienced at the hands of his father. He was not a man given to reflection; when he had a problem, he confronted it physically in work and play. He was full of himself and loved to tease people, but he was also generous to a fault. His younger siblings idolized him, a responsibility he took seriously. He set for them the example that he had been raised to think of as the key to survival: Be a man who works hard, who can laugh at himself, but who can defend his family. Be a rule breaker.

There was a softer side to Brian, too. While he was still away at school, he wrote to Moe, apologizing for how wild he had been when he lived in Tioga.

I don't want you and Mother to figure that it was your fault, he wrote, blaming his antics on small-town boredom and vowing that he had settled down. *I still think your [*sic*] the best father a son can have.* He closed by asking Moe to keep the letter to himself—*kind of a father to son letter.*

His longtime girlfriend, Debbie Thompson, adored him. They went to their first movie together—*The Sound of Music*—in elementary school, but didn't start dating until tenth grade. After that, they were inseparable well past graduation; they were homecoming king and queen and the toast of several proms.

If he'd asked Debbie to marry him back then, she would have. Like everyone else in Tioga, she loved the DeMoes, and she felt as though she were one of them. Instead, he left for Wahpeton. When he met new girls there, Debbie's eighteen-year-old heart was broken.

"I wanted to be in that family so badly," she said. Walking in through the DeMoes' back door was like coming home: There was always someone to pull up a chair, ready to concoct a plan to have fun, share a cup of coffee, tell an interesting story. Nobody ever seemed to be lonely inside those walls; to be inside was to be included, to be asked about your opinion, to be shown the latest project someone was working on. Even when the siblings squabbled, it was clear that they were watching out for one another, and they were devoted to Gail. They were effortlessly unpretentious. Gail could tell a bawdy joke as well as any of her boys; but she could also lend a sympathetic ear, and little children simply adored her. Debbie especially loved Gail. They stayed in touch for years after the breakup.

Karla always loved the warmth of that nest, but when her father went to Jamestown, she was finally ready to leave it. She was grown now, and she felt as though she needed to move beyond Tioga. She followed her boyfriend to Fargo—a comparatively large city on the other side of the state, and to her, the other side of the world. It was time to stretch her boundaries; her mother was physically safe now, even if she was emotionally shaky. The boys would look out for her, and Karla herself was only a phone call away.

In many ways, Karla's relationship with her future husband, Matt Hornstein, would echo Gail's romance with Moe. They met at a dance in the eighth grade; Matt was a newcomer, which immediately caught Karla's attention. His mother had died when he was fourteen, and he moved to Tioga to live with his aunt and uncle. Being the new kid gave him a certain currency in the little oil patch, where faces could become familiar to the point of monotony. He and Karla began dating, and soon they were an item. When Matt ran the projectors at his uncle's movie theater, a short walk from the DeMoe house, Karla sometimes went up to see him and kiss in the darkness—a fact that still makes her blush. She brought him home to meet her family.

Matt remembered Moe a bit more fondly than most; it was in Matt's nature to be generous, and he was an easygoing, reasonable person. "I was scared to death of him, of course, because I was dating his daughter. But he always liked me," Matt said. He thought of Moe as a quiet man who kept a nice yard, but he could see that Karla's home life was difficult; she didn't like being home if she could help it. He wanted to ease her burden.

Because Matt had watched his mother die from cancer, he understood, on a primal level, the loss Karla was feeling as her father deteriorated. It was difficult for Karla to talk about, and sometimes, she cried.

"I don't know what draws people together, but it's almost this magnetic force," Matt said. "You've both been through that, and you both try to escape it."

Things weren't always perfect for Karla and Matt. His agreeable nature was foreign to her, being so used as she was to her father's irate rants. When they were in high school, she very publicly broke up with Matt in the school hallway, throwing his class ring back in his face in front of several other people.

"I don't want to date you anymore!" she shouted. "You're too *nice*!"

"It was traumatizing," he said, laughing in hindsight. The split was temporary. Karla had been raised on the tale of her parents' young love that lasted, but the tragic way their story was turning out left her wary of happy endings. In time, she would learn to trust Matt, to believe that they could build a life together. And Matt, motherless child that he once was, found in the DeMoes a place where he could grab a toehold on family life while filling the gaps in theirs.

He worked as a drummer in a local band, harboring dreams of becoming a rock star. After graduation, he headed east to Fargo, Karla in tow. Four years later, they got married in Lake Tahoe, where a friend loaned them the use of a time-share property: he in a white satin-trimmed tuxedo, droopy mustache, and Allman Brothers hair, she in her best friend's high-necked lace wedding dress.

Gail and Doug accompanied them, and Doug—thrilled to be taking his first plane ride—gave his big sister away in the absence of their father. In 1983, it looked as though the young couple might have found the normalcy they'd both sought.

If Karla wanted to visit Moe, St. Cloud was only about three hours away. But like her siblings, Karla was at best ambivalent toward her father. Gail always defended Moe, saying the kids just didn't remember what he was like before the disease. But it was impossible to forget those terrible episodes when even her wild brothers were hiding from him, and their mother was running across the street to safety.

Occasionally, Karla had glimpses into a softer side of her father that hinted at some truth in Gail's claims. On one visit to St. Cloud, Karla and Gail took Moe on an outing to a mall. Spying a group of small children, Moe stopped in his tracks, excitedly waving his hands and trying to speak. It came out as gibberish.

Karla worried that Moe was scaring the children and their parents, but Gail explained that he thought the kids were his own: "He's going back in time," she explained. Such memory distortions are common in Alzheimer's patients.

In truth, Moe's absence—though it undoubtedly left her physically safer—left a void that Gail could never fill. Flirtatious by nature, she went on a few dates, even had boyfriends, but nobody quite approved of her having a romantic life when, technically, she was still married. She called and visited Moe, but she had no idea how long he would linger in his twilight world, or what he remembered from their lives together. It was the same cruel limbo that many spouses of Alzheimer's patients endure, but Gail didn't know that, or them, so she was navigating her new reality as best she could, trying not to worry about disapproving eyes.

She stayed in touch with Moe's family. His sister, Pat, had raised three daughters before divorcing her husband. Though they all lived near one another in the small towns that dot northern Wisconsin along Lake Superior, they were not close the way the DeMoes were. They bickered and often went for long spells without speaking to one another. Gail also busied herself with her children, keeping on top of their daily lives. She spoke often to Karla, who was now expecting a baby of her own, and she still had young Jamie to raise.

The DeMoe brothers lingered in Tioga longer than Karla and Lori, and the town shaped them as they moved into young adulthood. When Brian was twenty-two, he went to a party—as kids in Tioga have done for generations—at the Tioga Dam, where he met seventeen-year-old Christy

Thorson. They dated for a few months, and things weren't perfect—but instead of breaking up, they got married after Christy discovered she was pregnant. Like his father before him, Brian went to work in the oil fields, where he, too, would become known as an exceptionally hard worker. The other roughnecks nicknamed him "Toby Tyler, King of the Oil Patch."

Their first child, a girl, was born in 1978. Brian and Christy's second child was a boy.

All told, they were together for ten years, but many of those years were miserable. As had happened with Moe, Brian's personality was changing, though also like Moe, the source of those changes would always be clouded by his substance abuse. He verbally berated Christy. He became aggressive, often taking out his anger on the family pets, and his drug habit expanded; an unabashed pot smoker throughout their courtship, he graduated to cocaine. They both worked hard, but the family was broke. One Christmas, all they had to give the kids was a checkerboard.

But there was also a side to Brian that endeared him to people, especially Gail. He had a way with women her age; he doted on them, danced with them, paid them compliments and brought them little mementoes, sang songs they'd loved when they were younger—"You Are My Sunshine" was a favorite. His work ethic, the DeMoe hallmark, earned him a spot on a crew that went to Alaska to help with the cleanup of the *Exxon Valdez* oil spill.

The marriage finally breathed its last in 1987, and the children left with Christy. Later, during a rough patch in high school, Brian's son moved in with Gail and his uncles.

"Those were some of the best days," Gail recalled. Karla and Lori were gone, but her sons and grandson remained.

"It was always Mom and the boys," Karla observed.

Other families drifted apart as the children grew up and moved away, keeping in touch with a phone call every few weeks or a holiday visit once a year. But the boys were reluctant to leave Gail's house, returning to her often, protecting her in a way they couldn't when their father had lived with them. Doug, in particular, could never leave her. He never lived anywhere except Tioga, and he returned to live with Gail every time a relationship soured.

Karla and Lori copied her nurturing style with their own children, referring to Gail's example often. She had never been a woman of great means, but she made certain her door was always open, especially in times of crisis. For she knew those connections would help them survive, and the family she created with Moe would continue, even though his illness had taken him away.

Ironically, while the children's interest in Moe declined, doctors in St. Cloud were taking a closer look at his unusually young case of dementia, beginning almost as soon as he was hospitalized. His children knew nothing about it. A typewritten consent form dated April 5, 1979, described a project that aimed to study "mental confusion, disorientation, irritability and rapidly changing mood." The fifteen-week study planned to use three drugs: Hydergine, which was thought to increase circulation and activity in the brain; an antipsychotic drug sometimes prescribed for schizophrenics called thiothixene (which the consent form referred to as a tranquilizer); and an antidepressant called desipramine. Some patients would get a placebo.

Hydergine, an extract of a fungus that grows on rye, was the first drug believed to show some promise in treating Alzheimer's. A year after UC San Diego neurologist Robert Katzman wrote his editorial calling for an increase in research, the FDA approved its use in Alzheimer's patients sixty and older in 1977, two years before Moe began taking it.

The drug was developed by Albert Hofmann, a Swiss chemist best known as the inventor of LSD. Although it was originally created to treat hypertension, because it dilates blood vessels in the brain, it wasn't very effective at lowering blood pressure. However, some patients reported that it did improve their memory and mood, which led to its use as a treatment for dementia. Because of his age, Moe represented an opportunity to test Hydergine's effectiveness in younger patients.

"I understand that the reports and data collected from this investigation may not directly benefit me in the treatment of my disorder, but it is hoped that this study will provide information to benefit future patients with problems similar to mine," the consent form read.

The study, one of many run through the research service at the Veteran's Administration (VA) in St. Cloud, followed people with "chronic organic brain syndrome," a catch-all phrase used to describe any decrease in mental function for reasons other than psychiatric illness. In addition to Alzheimer's patients like Moe, it tested anyone with similar symptoms.

Underneath, in shaky penmanship that bore no resemblance to the letters he'd once written while courting Gail, Galen DeMoe scratched out his signature.

The study enrolled patients from nine VA facilities from August 1, 1978, through October 31, 1979. It turned out to be a bust. Though researchers hoped to enroll six hundred people, they only managed to recruit ninety-seven. The last patient follow-up was at the end of February 1980. No results were ever published, and a request to extend the study by three more years was denied. Scientists would continue to study Hydergine for two more decades, with most concluding that its benefits were extremely modest at best, though it did seem to help people with vascular dementia, which mimics Alzheimer's symptoms.

And thus marked the DeMoe family's first foray into research about the disease they did not yet fully understand.

Six

THE GHOSTS OF ANOKA

ONE AFTERNOON IN Fargo, North Dakota, when her father had been in the VA hospital for a few years, Karla sat across her kitchen table from a kindly nurse from the University of Minnesota named June White. It was the shortest of visits—just long enough for Karla to offer a sample of her blood—but in hindsight, June's work would prove to be an important stepping-stone in Alzheimer's research.

In Fargo, Karla and her husband, Matt, had settled in comfortably to family life. By 1985, they had two children. Perhaps because of her own difficult childhood, Karla was gentle in parenting her children. She wanted better for her kids, for them to avoid the embarrassment she had endured. They were Karla's universe, though the pull of her extended family—three hundred miles away in Tioga—was strong, much stronger than it had been when she was growing up. She spoke with Gail nearly every day.

June White had contacted Gail as Moe's next of kin for help with a study that June's boss was conducting on Alzheimer's patients. She had stopped in Tioga to collect similar blood samples from the DeMoes who were still living there, before heading across the state to Fargo to do the same for Karla.

At that point, though the family knew from his original 1973 diagnosis that Moe had Alzheimer's, they believed it was an extremely rare disease. Like much of the public, they hadn't heard of the 1968 British study linking Alzheimer's to the common dementia of the elderly. The researchers' latest interest seemed to reinforce their misconception: It made sense that they would notice Moe because his condition was so unusual. Though Karla knew her grandmother also had the disease, none of the children had interacted with Wanda much. On the few occasions when they had seen her, she seemed to them to be a very old woman. It didn't really occur to them that she had been much younger than her appearance and behavior suggested.

Karla was flattered by the researchers' attention. She felt important, maybe even a little bit famous. Getting blood drawn seemed to be a way to help science discover something new.

"It was a brief visit, I remember," Karla said. "I was excited that somebody was interested. . . . I felt that things would be better."

June White appeared to be a nice woman who promised to keep the family updated.

Despite the earlier warning of a possible hereditary link from the doctor who diagnosed Moe, neither Karla nor her siblings worried much about getting the disease. It seemed far too remote a possibility, and they were unaware of any of the breakthroughs in Alzheimer's research that were happening in the field, such as Foncin's and St. George-Hyslop's detective work in tracing the mutation that was plaguing Family N. They thought, after their difficult childhood days were behind them, that brighter days were ahead; the future seemed ripe with possibility. The fact that a research team was interested in them seemed just another chapter in their family's unique story, and they were happy to help, as Gail had taught them to be.

A mother of four, June White began her career as a registered nurse. But her true calling didn't find her until she started collecting brains.

She hadn't really taken to traditional nursing, so she accepted a job working in histology—the microscopic analysis of organ tissues—at Anoka State Hospital in Anoka, Minnesota, a suburb of the Twin Cities.

Built in 1900, the state hospital was a redbrick sanitarium constructed over a maze of tunnels said to be haunted by the ghosts of those who died there. Beginning in the 1950s, Anoka routinely collected brains from patients at all of Minnesota's state institutions; when a patient died, the body was autopsied and the brain was separately shipped to Anoka for a neuropathologist's diagnosis. June's job was to serve as the brains' curator. The gray matter arrived in all sorts of containers, including, memorably, no. 10 cans of the variety used in restaurants for bulk supplies like stewed tomatoes; they all landed in June's collection. By the time she met Leonard Heston in 1970, she had stockpiled more than two thousand brains.

Heston, a psychiatrist and geneticist who had newly transferred to the University of Minnesota, was interested in the origins of Alzheimer's disease. Like a handful of other scientists who were studying Alzheimer's, he suspected that genetics were a factor. But to test his hunch, he needed brain tissue from afflicted patients, which was hard to come by.

He was thrilled to hear of the repository at Anoka, and for about a year, he traveled there regularly, where he read through each chart and searched for tissue with plaques and tangles. As has often been the case with Alzheimer's, many of those patients were misdiagnosed or lumped together under the heading of "senile dementia"; Heston's more practiced eye found Parkinson's disease, Pick's disease, and about sixty families with Alzheimer's.

In 1973, a dearth of state funding closed June White's office. Before she left, she called Heston to offer him the brain collection. As she prepared to send him the slides and paraffin blocks of brain tissue, along with their accompanying files, Heston made her a proposal: If she would come to work for him, he could give her a raise and a four-day workweek.

"What will I do?" White asked him.

"That will be a surprise," he answered.

On her first day of work, June found her office stocked with the brains as well as the patients' death certificates. Heston asked her to use the documents to track down the deceased's surviving family members so she could collect their blood samples and help establish their pedigrees, which would be used in tracing a genetic link for the disease. It was a curiously personal process, one that brought scientists face-to-face with the people they hoped their work would one day help.

White became a self-styled detective, combing through phone books and newspaper obituaries to locate the elusive families, most of whom—like the DeMoes—had no idea what Alzheimer's disease was.

The pursuit of an answer to the riddle of Alzheimer's made White a seasoned traveler. She put thousands of miles on rental odometers; she flew to England; once, she hired a small plane to take her to a Canadian family in Regina, Saskatchewan. She drew blood samples from a family of thirteen children in Mexico, then casually toted the vials back across the US border in a briefcase. She put one brain through a luggage security scanner at an airport; another, packed in a Tupperware container, was handed to her by a doctor who met her in a hotel parking lot. Her experience echoed that of the search for members of Family N; many genetic researchers spent countless hours wearing out shoe leather in search of affected relatives. Their knowledge was hard-won, and for some, the personal relationships they developed with the families would create a lasting emotional bond. For others, it was easier to move on.

In 1974, Heston published his first paper discussing a connection between the disease and genetics, in *Science*. White was listed as a coauthor. They continued to write papers together well into the 1990s.

She got to know a few of the families she studied and did her best to follow up with them, but eventually, the University of Minnesota's participation in the study also petered out. Alzheimer's was, at the time, still a drastically underestimated disease; as such, funding remained relatively thin. By the early 1990s, Heston and White had sent their sample collection to the University of Washington, where Heston eventually transferred; a bigger sample size increased the chances of finding more definitive answers in the genes of the afflicted.

• • •

Meanwhile, when a much-deteriorated Moe was transferred out of the VA hospital in January 1986, he was moved to the Americana Nursing Home in Minot, North Dakota, about eighty miles east of Tioga. Although he had managed to recognize his family longer than they'd expected, his final decline had begun, opening fresh wounds for the DeMoes. Now that he was physically closer and significantly weaker, they were reminded of his existence in ways they hadn't been—except for Gail—in years. He hadn't been their dad for so long. Karla hated to admit it, hated how it sounded, but they'd largely forgotten him. Most of the time, he slept. He had entered St. Cloud a pudgy 214 pounds. By the time he left for the nursing home, he had withered away to 134 pounds. All his meals were pureed in a blender.

In the spring of 1989 the nursing home called to say Moe, now fifty-eight, had developed a high fever. Gail and the kids who had remained in North Dakota—Karla, Doug, and Jamie—went to keep vigil. The staff weren't sure what had caused the fever; they thought he might also have suffered an aneurysm in his brain stem. Moe's temperature finally returned to normal, but he never really recovered from the incident. His breathing was labored.

"He lost so much weight—there was just nothing left of him," Karla said. "And the IV wouldn't stay in anymore."

Gail stayed at his bedside, rubbing his neck, sleeping in a chair in his room. None of their friends came to see them, except for Hank Lautenschlager's daughter, Roxanne. Moe and Gail, once the most popular couple in town, were on their own at the end. Gail was sure Tioga had forgotten them.

Shaped by those experiences, Gail would never judge anyone. She openly accepted anyone who sought her counsel; there was not a bad marriage, an unruly child, an abusive parent, or an inner demon that would nullify her friendship.

When the call came about Moe, Karla knew she had to be there for her dad, even if he hadn't done a good job of being there for her. Her husband, Matt, had to stay in Fargo to work, so Karla left their children—who had never really known Moe—with a friend while she waited at the nursing home with her family. She kept hoping someone, anyone, would come to their rescue and help them decide what to do next. Should they agree to a feeding tube, or let him die? If they did let him die, was that wrong? Her father had lived so long with every ingredient of his personality gone; was it humane to step away? Nobody offered any advice or support.

Eventually, the nursing home sent them back to Gail's house to rest.

"We have to start cooking," Gail told Karla. Brian, Lori, and Dean, who all had moved out of state, would be home for the funeral. Since nobody came to see them during their vigil at the nursing home, Gail did not expect anyone to show up at a funeral, either, bearing the customary casseroles and dishes that people often bring to bereaved families.

They were back in Tioga for a day when they got another call: Come right away. This is the end.

The trip took an hour and a half. When they arrived, the nursing home administrator informed them that by law, they had to try to put the IV in again, because of how dehydrated Moe had become.

"Oh my God, he's almost dead, and you're going to bring him back to life?" Karla thought. She started crying. It was as though Moe's illness would wring from her every last drop of strength she had.

Whatever the final efforts were, they failed. A nurse returned to the family to say that Moe was gone. Though a priest had read him his last rites the day before, neither his wife nor any of his children had been with him when he died.

When they arrived back in Tioga, the rest of the siblings had shown up. They passed the time by sitting around and reminiscing, and the out-of-towners teased Doug for being such a homebody, since he had, once again, moved back in with their mother. It infuriated Karla. Her gentle brother, who had borne so much of their father's abuse, had still gone to Moe's bedside, supporting Gail in these final days. So what if he still lived in Tioga? Who were they to judge him?

A year earlier, during a rough patch in her marriage, Karla had gone to see a counselor, who asked her to write down some good memories of her childhood and bring it in the following week. Karla couldn't remember any; it was as though they'd all been wiped clean.

As she sat in her mother's house, preparing for the funeral, Karla leafed through some old photo albums with her siblings. There were the memories she'd forgotten: Moe's brother, their uncle Jerry, who came to live with them. The garden hose that Brian had used to chase their mother through the house. The dances, the holidays. Her anger started to dissipate.

As they thumbed through the pages, she let the rekindled joy of those happier moments seep back in. She looked around at her siblings, and she felt the counterweight of their collective resentment against their late father start to ease up, ever so slightly. They were the only ones who knew just how bad it had been, how terrifying he could be. They had survived him together; it was a bond that sustained them. Now that they were at the end of that road, Karla felt, for the first time, how strong they actually had become.

Gail's door banged open. Neighbors started showing up with covered dishes, dinners, tempting desserts—enough food to feed a small army, more than the family could ever possibly eat. The town would turn out, in droves, for Moe's funeral, many of his old friends—who had not seen him since he left Tioga in a police cruiser ten years earlier—expressing shock at how shrunken his body was.

The DeMoes had not been forgotten. The town did not hate them. It was just that nobody had known what to say.

The autopsy report was dated June 9, 1989, and the consulting psychiatrist was Dr. Leonard Heston from the University of Minnesota, June White's boss.

According to the autopsy, before Moe's brain had been soaked in formaldehyde, it weighed 1,050 grams, or about 2.3 pounds; a normal adult human brain weighs about 3 pounds. Both hemispheres had shrunk, and damage to the frontal cortex—which controls emotions, problem solving, reasoning, planning, and other ingredients in personality—was particularly severe. So, too, was the insula, which is a prune-sized section of the brain that controls cravings, self-awareness, and social emotions such as empathy, guilt, and disgust. All were common patterns of distribution in Alzheimer's.

Under a microscope, stained sections of Moe's brain revealed "enormous numbers" of plaques and tangles. In the sample from his left calcarine cortex, which controls visual perception, there were numerous plaques, frequently with well-developed amyloid cores.

The report concluded that Moe's brain showed severe changes "in keeping with the presenile onset of his clinical picture of dementia, which is attributable to Alzheimer's disease." At last, the autopsy confirmed what the doctor in Fargo had said years before: Moe's strange behavior was, indeed, caused by the same disease that had killed his mother.

The search continued for concrete genetic links that bore out the pattern of autosomal dominant inheritance that Robert Feldman and Jean-François Foncin were exploring in Family N. Though Dmitry Goldgaber had discovered the APP mutation on chromosome 21, the initial excitement that followed was short-lived. In families known to carry this inherited form of Alzheimer's, it turned out that only a tiny percentage tested positive for APP. Family N, for example, did not. So now, a new gene-cloning hunt was under way, searching for other possible autosomal dominant mutations, even when Moe was still withering away in the hospital.

Peter St. George-Hyslop, the British neurologist and molecular geneticist who had teamed up with Foncin in 1985 to study Family N, had moved from Boston to the University of Toronto. Fortified in part by the access to Family N, he and his team would continue their search for an additional mutation through 1995, adding families as they found them. Other, competing teams backed both by industry and the National Institutes of Health were also after the same prize.

"It was an exciting project," Hyslop says. "If you could find the gene, you could understand more about the disease, which at the time was a huge enigma." Finding the gene meant possibly being able to predict the disease, but it also would allow science to understand the molecular changes that occur as Alzheimer's unfolds, which in turn would create better drug targets.

It was difficult work, often riddled with frustration and disappointment. At first, they weren't even searching in the right neighborhood. Hyslop and his collaborators initially thought chromosome 21—where Goldgaber had identified the APP gene—would also be the hiding place for the Family N mutation.

One of the teams competing with Hyslop's was located at the University of Washington, where June White and Leonard Heston's enormous brain collection, including the DNA donated by the DeMoes, now resided and Heston now worked. The DNA analysis of the brains was in the hands of a bearded biochemist named Jerry Schellenberg.

Like the others, including Hyslop, Schellenberg's group had been collecting DNA samples from families who seemed to be inheriting early-onset Alzheimer's from one parent—either male or female. Each child was born with a fifty-fifty chance of getting the disease in middle age.

Because autosomal dominant mutations are so rare—occurring in only 1 to 5 percent of all Alzheimer's patients in the world—those odds work against finding families with the disorder; some of it is luck, and some of it comes from old-fashioned research. Schellenberg, like Hyslop, benefited by connecting with another scientist who had a large collection he was willing to share.

In 1991, Schellenberg naturally had no genome map to use as a reference guide. So instead, he worked to identify the mutated gene by comparing pieces of DNA from afflicted families—known as "markers"—to markers that other scientists were associating with Alzheimer's disease, looking for a match that would point to the chromosome with the faulty gene.

"We were literally plucking markers out of the air," he recalled.

It was like finding yourself in New York City without a map and looking for a specific address. If you went down every street in the city, and were systematic about the way you did it, you'd eventually find what you were looking for; it was only a matter of time.

"The first time you do this, you're really excited," Schellenberg explained; it was like buying a lottery ticket. The world seems full of possibility, the odds of winning huge. But after hundreds of failed attempts, the anticipation of a scientific eureka fades.

Schellenberg's job was to align the lab work that one group on the team had completed with the statistical analysis of another. They'd tried something in the neighborhood of 150 genetic markers, focusing on chromosome 21 and more recently branching out to chromosome 14. Each failure represented weeks' worth of work by about 140 people; messy, tedious work that had to be scored, analyzed, and categorized by family before revealing itself to be worthless.

In June 1992, eighteen months into the gene testing, Schellenberg was waiting for a printout of the numbers associated with a genetic marker on chromosome 14, expecting the usual negative numbers. But when he looked down, positive numbers were spewing out of his printer.

"I thought the key that hits the minus sign must have been broken or something," he recalled. Instead, he had struck pay dirt: He now knew he was looking in the right neighborhood, making his search much, much narrower—more like searching a city block for the right address instead of the entire metropolis.

The marker matched the DNA of so many of his Alzheimer's families that Schellenberg was certain he'd find, somewhere on chromosome 14's sequence, the mutation responsible for the vast majority of inherited cases. Unlike APP, which only matched a small percentage of affected families, this one would explain Family N and many more, including the DeMoes.

Before the paper with the Schellenberg team's findings could be published, an international meeting of Alzheimer's researchers was scheduled in Italy. Worried that his team might get scooped before its discovery appeared in print, Schellenberg went to the meeting with a slide that had their results on it. If someone else tried to present the same information, he wanted to be sure he could prove that they were at least tied for first.

As soon as he realized the discovery was still theirs alone, he went home early. The next goal: finding the specific mutation on chromosome 14—that mystery address he'd been hunting—that was the source of the matching marker.

Three years after Moe's death, Karla received a note on University of Minnesota letterhead:

Oct. 20, 1992

Dear Ms. Hornstein:

In the October 23rd issue of Science *magazine, an article will appear reporting the definite linkage of a gene causing Alzheimer's disease to a small segment of human chromosome 14 in some families.*

There are no immediate consequences of this to you and your family. However, the two molecular genetics laboratories involved in the study (at the Universities of Minnesota and Washington) are continuing to work toward the specific location of the gene and no doubt other laboratories will soon join them. This means that it may soon be possible to predict on the basis of a blood test very early in life who will and who will not get the disease. Not everyone will want this information but it is time to start thinking through the choices that may soon be presented to you.

In the longer run, this is a most important breakthrough. After the gene is isolated, the goal will be to discover what it does or fails to do that results in Alzheimer's disease. Then ways to compensate for this error will be sought—a drug perhaps, or maybe only an adjustment of diet. It may even become possible to replace the faulty gene. The bottom line is that knowing the structure of the gene will provide rational bases for preventing the disease.

I realize that you have waited for this very good news through long trials of fire and heartbreak. Thank you for your wonderful cooperation. It has made a great advance possible.

Sincerely,

Leonard L. Heston, M.D.
June A. White

The letter came with a copy of the *Science* article, authored by Schellenberg. Coauthors included Leonard Heston and June White. An accompanying article said the team had been studying DNA sequences in nine families with a hereditary pattern of early-onset Alzheimer's, though it did not specify—and Schellenberg could not say for certain—whether the DeMoes were one of them. All seemed to carry a similar genetic marker on chromosome 14.

Karla wasn't really sure what any of it meant, except that maybe it might be a way to find out if someone would get the disease earlier in life. She had taken an administrative job at a small college, so she went to the library at work and tried to look up more information, but the article was too technical to make much sense to her. It was the last time she ever heard from June White, who retired three years later.

In hindsight, the letter should have served as a warning bell for what was to come. Karla would save the letter and the article, but another decade would pass before she would read it again with any sense of urgency.

Despite Schellenberg's triumph, it was Hyslop and his team who finally identified the mutation in 1995. He named it PS1, or presenilin 1. It was the most common mutation found in this rare autosomal dominant form of Alzheimer's; it affected the DeMoes, Family N, even Auguste Deter, the original patient identified by Alois Alzheimer.

Of the many teams competing for the mutation's discovery, some backed by financial resources such as the NIH, Hyslop's was decidedly the underdog. He and his partners, Foncin and an Italian, Amalia Bruni, were working with the research equivalent of pocket change—a little more than $100,000 per year. Bruni, in fact, was pursuing Family N as a volunteer in her spare time.

Yet Hyslop thought their lack of financial resources might have had an unexpected benefit. Without the luxury of funding, they couldn't afford to chase potentially fruitless avenues. "Sometimes, when you don't have enough money, you have to think out very carefully what you're going to do, and conduct the right experiment," he said.

And luck, he added: It helps to have a little of that, too.

A few family members the group had studied remained in contact with Hyslop. One in particular, a pathologist, called monthly, offering to help and asking for updates on the progress the team had made. He had been so worried about passing the disease along to a new generation that he had deliberately avoided having children, and had set a life goal of convincing someone to find a solution.

About six months before they found the gene with the mutation, he stopped calling; he'd fallen ill.

Hyslop waited until he was absolutely certain they'd identified the gene; then he called the man to report the news. Although he couldn't come to the phone, the man's wife relayed the news. "He is intensely grateful," she told Hyslop in that call. "It's something that he's been waiting for for a long time."

Later that night, he died from a malady unrelated to Alzheimer's disease. He hadn't been a carrier.

Seven

UNTAMED HEARTS

NOW THAT MOE was gone, fifty-four-year-old Gail was single for the first time, really, since she was twelve years old. All her children were essentially grown; even Jamie had turned eighteen. It was time to move on with her life.

In the aftermath of Moe's 1989 funeral, the town's judgment had softened. There was no shortage of gentleman callers for Gail DeMoe, known throughout the town as Grandma Gail, who believed in living life to the fullest and practiced what she preached. "Grandma had her own secrets," Karla says. "Mother was quite the partier in her day."

When they were out in public, if she saw a man close to her age, Gail would joke with her kids: "That could be your new daddy!"

Her message was simple: Do not waste the present by dwelling on the past. But she never forgot Moe. He would always be her husband.

Of Moe and Gail's six children, only half—Brian, Doug, and later, Jamie—lingered in Tioga to devote the better part of their working years to the oil fields. Although Brian ventured briefly elsewhere—he remained "Toby Tyler, King of the Oil Patch," unrepentant pot smoker and respected veteran of the local bar scene.

Doug, the next oldest boy, stayed rooted in their hometown his entire life. His comfort zone extended about as far as the town limits, and sometimes not even that far. Doug worked for the same oil company from as far back as anyone can remember, even when competing firms came to town. He worked only one job—latching pipes on drilling rigs. Everyone in the oil patch has a nickname: Toby Tyler, Chrome Dome, Belt Buckle, Weiner. Doug's was Doug Latch. Even his children called him that on occasion.

As they grew into adulthood, Doug and Dean no longer spent as much time together. Doug's closest high-school friend, Gary Anderson, had also moved away, so Doug often found himself at loose ends.

Without Dean, he seemed incomplete. He filled the void left by his brother's absence with the companionship of the women he dated, becoming dependent on them, and—increasingly—on Gail.

Shortly before Karla and Matt got married, Doug became engaged to his then girlfriend, a woman named Lola. It was a tumultuous relationship, and Doug was needy and nervous. She broke off the engagement and returned the ring. The DeMoe siblings chipped in to buy it from Doug, then gave it to Matt for Karla.

Eventually, Doug and Lola reunited and married when Lola got pregnant with their daughter, Jennifer. The marriage ended when Jennifer was three, and for the next decade, Jennifer saw her father only on holidays or over the summer. Even so, she always thought of them as close. She had a little half brother, Brady Thompson, but never felt as close to Brady as she did to their dad.

At thirteen, around the time Brady was born, Jennifer went to live with Doug at Grandma Gail's house, but he hadn't had much practice as a parent. Though she grew into a pretty woman with a husky voice and a booming, hearty laugh, Jennifer was a chubby teenager, and Doug's strategy for addressing the issue was to lock up food.

Jennifer lived in Minot for a while, and traveled to Montana sometimes with her boyfriend. But her father stayed behind. He didn't take vacations; he didn't have hobbies; he didn't know how to use a computer. His only real interest was his love of new vehicles: trucks, motorcycles, power boats—anything with a motor.

"His life was this town. He had no need to see what else was out there," Jennifer said. "It's always been a safety blanket.

"Brian, my dad: They're very simple men," she observed.

At the opposite end of the spectrum was Lori. Of the DeMoe siblings who left Tioga, Lori was easily the most widely traveled. In the nearly four decades Steve McIntyre spent working on the railroad, he traveled to every state west of the Mississippi, from New Orleans to Seattle, Chicago to Los Angeles. For a significant chunk of that time, Lori was with him—first in their travel trailer, moving from one city to the next while the crew worked on the tracks. When it was time to move to the next location, they simply tied their refrigerator and cupboard doors shut, moved loose objects to the floor, and jumped behind the wheel. The gypsy lifestyle suited Lori's restless spirit perfectly; on the road, and in Steve, she found a contentment she had never known.

One weekend in 1979, Lori told Steve she was pregnant. Their wedding plans abruptly shot to the top of their priority list.

Even "when I was young, I knew that I didn't want a big wedding," Lori said. "I always knew that. We had a redneck wedding," she added, laughing.

Their first daughter arrived on August 3, 1980, and a second baby girl followed three years later. For about a year, the young family lived in the twenty-eight feet of space afforded by the travel trailer. Then they upgraded and moved into a bunk car, which was like a trailer house on wheels, equipped with a dining nook, a bedroom area, a bathroom, and a shower.

Lori and Steve would be married five years before they moved into their first house, in Kimball, Nebraska. They loved their wandering life, never knowing where they'd wind up next. They were about as far from the monotony of Tioga as they could get.

One fellow railroad wife remembers Lori cheerfully hauling clothes to the Laundromat, baby in tow; she was always working on a craft, usually an afghan or a baby blanket for one of her friends. She was busy, capable, unsinkable.

Though he was traveling during the week, Steve's heart remained with his family, and his admiration for the masterful job Lori did was apparent even when he was frustrated with her. "She *raised* those girls," he would say. As they grew, the girls became small reflections of their mother: plucky, fiercely independent, creative, and fearless.

Steve and Lori's third daughter was born the year they moved to Kimball; one of her nurses, Robin Tjosvold, would become one of Lori's closest friends for the rest of her life.

During those years, Lori's independence served her well. With Steve home only one day a week, she never missed a beat; her seemingly limitless energy was enough to raise three active girls and take care of other people's children, too, in a de facto backyard day care. Every birthday cake the girls ever had was home-baked and decorated by their mother, and they were the best cakes in town. She sewed all their Halloween costumes and a lot of their clothes, as well as clothes for their dolls. She was always deeply involved with them—coaching a competitive jump-roping team, taking them camping, dancing, listening to their triumphs and heartbreaks.

"She redecorated every house we ever lived in," her daughter remembered fondly. "Everything on the walls was handmade by Lori; every Christmas ornament on the tree."

Lori took care of the odd jobs around the house, too, tending to the lawn and functioning as a self-taught handyman.

Eventually, Steve earned a promotion that would move the family to Laramie, Wyoming, and allow him to be home with his girls every night. Cradled by the Laramie and Snowy mountain ranges, it was the kind of place that appealed to the McIntyres' free-spirited family, and for them, it became a hometown.

Lori wasn't particularly close to her younger brother Dean while they were growing up, the way Doug had been. But in adulthood, the similarities in Lori's and Dean's personalities became more apparent, and they grew closer, especially once they started raising children. Ask anyone in the

DeMoe family to name their favorite sibling, favorite cousin, or favorite uncle, and invariably, one name came back: Dean.

He was always the wittiest of the four boys, fun-loving and handsome, with laughing eyes, a dimpled chin, and a shock of ruffled dark brown hair. Compactly athletic, he was a study in contrasts: Eventually he became a dedicated family man with more than a touch of a wild streak, widely respected in his business dealings, but just reckless enough to be dangerous. Even in a family of men admired for their work ethic, he stood out. Dean regularly did the jobs of seven men while working for an asbestos abatement company. In his forties, when he returned to the oil fields, he frequently put in seventeen-hour workdays, shaming the effort of men half his age.

Like Doug, Dean fathered a daughter with a girlfriend, but unlike Doug, he resisted settling down. He sometimes jokingly referred to weddings as "another funeral," and he vowed he would never marry. But he loved his daughter and fought hard to stay in the little girl's life after his relationship with her mother ended. After initial custody struggles, Dean established a visitation schedule with his daughter, who had moved to Fargo with her mother and her stepfather.

Dean was a man who valued loyalty, especially in the face of potential disaster. He and his best friend, Monte Olson, met as juniors in high school while working summer jobs at a sprawling gas plant on Tioga's outskirts. Monte was driving a work truck back to the plant to pick up some tools, and Dean was loading sulfur with a backhoe. The backhoe's muffler somehow sparked, igniting a fire. At a gas plant, that was an instant emergency. In that split second, both boys—sons of the oil field—recognized the potential for disaster. Monte jumped out to help.

"We both didn't know shit about nothing," Monte said, "and we both, together, got this fire put out." It was an experience that would bond them and serve as a metaphor for the days to come: It seemed as though they were always operating just this side of danger, and they always managed to bail each other out. Over time, while Doug became more vulnerable, Dean began to feel invincible.

After high school, Monte went to college to study engineering, and Dean stayed on the rigs, but they still worked together from time to time, surviving at least one more near-miss industrial accident. The two men boxed together, drank together, and chased women.

The same year that his daughter was born, Dean finally met his match in the unlikely form of Deb Clark. She was six-foot-one, a standout basketball and volleyball player for Jamestown College. The daughter of a quiet, respectable insurance company man, Deb came from a devoutly Christian family; neither of her parents drank or cursed.

Dean wasn't intimidated by Deb's height, though when they began dating, he did develop a habit of walking on the curb to seem a few inches taller. On one early date he took her to a water park, where Dean tried to impress her by flinging himself down the tallest slide—backward. The park staff promptly kicked him out, a recurring theme throughout his life.

Deb's parents were less than thrilled with her new romantic interest.

"It's not that DeMoe boy from Tioga that boxes, is it?" her father asked. It was.

"I had to really convince my dad that maybe that would change," Deb remembered.

But change Dean did, and remarkably so. In September 1987, about a year after they met, a friend of Dean's moved to Colorado and took a job in asbestos removal. Dean joined him in Denver, and Deb, who was looking for a teaching position, followed along.

It was a relationship that would shape the two wildly different personalities from within. Dean mellowed and began adapting into a family man; Deb, the dedicated Christian, lived with him, becoming his common-law wife without the benefit of a church wedding until years after their two children were born. She struggled with that dichotomy, particularly since her parents did not approve of their arrangement. But such was her dedication to Dean that she waited him out until he finally decided to marry.

In 1989, their son was born.

In 1996, Deb gave birth to Dean's second daughter.

Despite how pleasant life was in Denver, Deb always imagined the family would wind up back in North Dakota. She missed her parents and both of their extended clans. Their chance came in 1997, when the Red River Flood wreaked havoc on the eastern side of the state, causing billions of dollars in damage around Grand Forks. A friend of Dean's won a four-year contract to remove asbestos in buildings that were damaged by the flood. He contacted Dean and offered him a job.

"We wrote a pro and con list out," Deb recalled. "Actually, there weren't many pros for leaving Colorado, other than family. But that outweighed everything else on the other side." Living in Grand Forks would put them close to Karla and Matt, and to Deb's sister.

In deciding to move back home, they also went ahead and fulfilled one of Deb's long-held wishes by finally having a church wedding on December 23, 1999, the same anniversary as Deb's parents. Every year thereafter, Dean and Deb went out to dinner on the anniversary with her parents, who had come to love Dean as a son.

Dean settled in, and, as was his habit, got to work. He traveled often for the asbestos removal company, calculating estimates and overseeing the actual job. But when he was home, they were happy; he loved playing pickup basketball in his driveway with his kids, who had inherited their parents' passion for sports. The loser had to make milk shakes, and Dean let his younger daughter win quite a bit.

Though most of the DeMoe siblings harbored bitter memories of their father, it was Jamie, the youngest, who lived deepest in his shadow. He closely resembled Moe; they had the same sandy hair and sincere smile, their eyes crinkling identically from long days spent outside.

Jamie remembered little about his father except the two things most people remember: how big he seemed to be and how hard he worked. "I never really got to hang out with him or anything, because he worked all the time," he said.

These memories most likely come more from his siblings than from Jamie's own experience, though, since Jamie, born in June 1971, was nine years younger than Dean, and just seven years old when Moe departed for

Jamestown. In the years leading up to that, when Moe was forced to quit working and Gail took a job to support the family, Jamie was stuck at home with his father, who was often not in his right mind. If Moe was asleep while Gail was at work, the little boy would just run around outside by himself.

Whether because of Moe's volatile omnipresence or the age gap between Jamie and the rest of the pack, Jamie's was not an easy childhood. The hearing in his left ear was somewhat compromised, and he struggled with learning disabilities that Tioga's small-town school system was slow to address.

At home, his brothers liked to tease him for his slight build and for being so much younger than everyone else.

"You're the mistake!" they'd tell him, even as an adult. Yet despite their ruthlessness with one another, their loyalty to those they loved was unwavering.

Sometimes the older kids would babysit him or take him to the movies. But Karla also admits guiltily that she ducked out on "Jamie duty" more than once. She remembered driving past her mother's house in a car full of friends, seeing her brother sitting on the curb, and announcing, "It's not my turn to watch him!" before zooming off.

Those early, fractious years began to take their toll when Jamie moved into junior high school. He started to misbehave. And Gail, weary veteran of her husband's explosive violence and the endless carnival of her five other children, was spent. In 1983, she sent him to briefly live with Matt and Karla, who was thirteen years older than Jamie, in Fargo. He attended junior high in the larger school district, which finally identified his learning problems.

He lasted in Fargo only a few months. Karla and Matt had their own baby and just one car between the two of them. Jamie could walk to school, but Matt and Karla both worked, and by the end of the day, they were exhausted. They'd come home to find that Jamie had set up a Monopoly game, hoping they'd play with him, but they were just too tired to oblige. Eventually, they sent him back to Tioga; but there, finally, after years of being everyone's afterthought, he began to hit his stride. He didn't seem to resent his mother for sending him away, or his sister for sending him back. In fact, he would remain close to both of them as he grew into adulthood.

Most people, including Jamie himself, described him as a classic worrywart. An undercurrent of anxiety flowed through everything Jamie did, whether he was working, playing golf, or even driving his car. He checked over his vehicles meticulously for any sign of malfunction, hoping to ward off disaster. In that way, he was like his brother Doug; he always expected the worst and prepared accordingly.

Despite his innate pessimism, in high school Jamie developed a reputation as a ladies' man. He was handsome and outgoing, and loved the usual Tioga pursuits: snowmobiling, tearing around on motorcycles, partying at the dam.

Jamie was only a few years older than Brian's children, and nine years older than his nephew, Brian's son. In some respects, they were closer to being his peers than his siblings were.

"He was the hottie all the girls were after," said Doug's daughter, who was thirteen years younger than her uncle Jamie. The family joke was that between Dean, Doug, and Jamie, the brothers had dated every available woman in Tioga.

Though he grew up immersed in his family's sense of humor, Jamie never really learned to roll with the punches. He hated being the butt of jokes. He was sensitive to how other people perceived him and defensive about his careful habits. If it took most people three minutes to accomplish a task, it would take Jamie twenty, because he wanted to think each step through carefully and then double-check it. In an effort to stay on top of jobs, he created to-do lists on Post-it notes that he stuck all over the house. He was zealous about the lawn, too, mowing it to perfection. But if a pipe burst or another household crisis suddenly erupted, he froze; adapting quickly to surprises simply wasn't in his nature.

Their individual idiosyncrasies notwithstanding, the DeMoes generally believed their fight with Alzheimer's was over once Moe was gone. The siblings packed away their memories of those hard years—including the letter Karla received from Leonard Heston and June White—the way Gail stored Moe's letters in a box under the bed they'd once shared. Their lives were now their own, they thought, and the future was theirs to shape, unfettered.

Part TWO

What's past is prologue.

—*William Shakespeare,* The Tempest

Eight

A BLAMELESS AND UPRIGHT MAN

Then the Lord said to Satan, "Have you considered my servant Job? There is no one on earth like him; he is blameless and upright, a man who fears God and shuns evil."

—Job 1:8

IN 1995, THE extended DeMoe family gathered for a reunion in Brule, Wisconsin, where Moe's sister, Pat, had settled with her husband, Rob Miller, and their three daughters. Traditionally they got together at Gail's place in Tioga over the Fourth of July. But that year, they all trekked back to Moe's home state and Pat's cabin—Gail, her six children, their spouses and families, and Pat's three daughters, who were also grown and raising children of their own.

Gail and Karla were particularly excited to see Jerry, Moe's younger brother who had spent his high-school years living with Gail and Moe after Wanda's death. Now Jerry was forty-nine years old and living in Oklahoma with his wife and daughters. Gail had been like his second mother, but these days she seldom got to spend time with him.

Perhaps it was because she hadn't seen Jerry in so long; perhaps it was because

Moe's decline and death six years earlier were still so vivid in her memory. But Gail noticed immediately the difference in her brother-in-law: The repetition. The halting speech. The blip of vacancy in his eyes. When he went golfing, he didn't know which way to hit the ball. Karla saw the pattern, too.

It was all much too familiar.

Gail pulled Jerry's wife, Sharon, aside at the reunion and urged her to face the problem. For Gail, better than anyone in her family, knew what hell lay ahead.

Though she didn't act on it right away, Gail's warning stuck with Sharon. While she knew Jerry's mother and oldest brother had died from Alzheimer's disease, she and Jerry had never talked about it much, or considered it relevant to their lives. She remembered disagreeing with the decision to put Moe in a nursing home, though it didn't change her affection for Gail and the rest of the family.

"In my mind, it wasn't going to happen to us. I thought Jerry was different," she said. "I don't know why."

Jerry and Sharon had a full and active life together; married in 1968, they had two girls. Jerry taught diesel mechanics at Oklahoma State University Institute of Technology. Sharon was a homemaker, and eventually, she homeschooled Sheryl, their younger daughter. They were deeply involved with their Christian church, and in their spare time, the family took food and necessities to the needy.

Jerry talked a lot about his longing for a boy in the family, but that never interfered with his adoration for his girls. When his oldest turned sixteen, he rebuilt her a red Volkswagen bug and taught her how to change its oil; for his second daughter, ten years younger, he forged a go-kart from scratch. He was also part owner in a plane and spent hours working toward his pilot's license.

In the wake of Gail's comment at the reunion, Sharon forced herself to pay more attention. The truth was, small lapses were starting to add up to a more ominous picture: Jerry, always a genius at anything mechanical, was beginning to struggle at work. The school was starting to emphasize technology as the diesel field became more computerized, and Jerry was having to ask some of his colleagues for help. Sharon minimized these struggles initially, but after both Gail and Karla pointed out changes in Jerry, her small anxiety began to grow.

Jerry agreed to get tested at the National Institute of Mental Health in Maryland, just to be on the safe side. Sharon sat Sheryl, her younger daughter, down to explain that her father might have the same disease that her uncle Moe had had. Sheryl took it in stride; she and her sister had been so sheltered by their parents' love growing up that she couldn't imagine things not turning out fine.

"I had no idea what was to come," Sheryl said.

She thought her father might be somewhat forgetful, or act out of character. But she had faith in their strong family bonds: "He would need our help, and he would eventually die. OK, everyone does, right? He's my dad, I love him, and I will do whatever I needed to do for him."

In addition to APP and PS1, a third mutation—PS2—had been identified six months earlier on chromosome 1. But blood tests were not yet commonly available, even to NIMH. In the absence of an actual, physical diagnosis, in Bethesda Jerry underwent two weeks' worth of tests such as an electroencephalogram, or EEG, in which electrodes attached to the head record the brain's electrical activity, and pencil-and-paper evaluations. The results pointed to the worst: Through process of elimination, he was diagnosed with "probable Alzheimer's disease." At forty-nine, Jerry had been declining for two years. He had had some hearing loss related to his military service and a history of migraine headaches, and he struggled with arithmetic, memory, and reading. His IQ was measured at 84—placing this bright, mechanically talented man in the low-average intelligence range.

Doctors prescribed Tylenol to help with his headaches and a low-cholesterol, high-carbohydrate diet rich in fruits and vegetables. The diet was widely promoted at the time as a heart-healthy option, and both obesity and cardiovascular disease have been associated with a greater risk for Alzheimer's. Because the heart pumps about 20 percent of the body's blood to the brain, any damage to the heart or blood vessels affects the brain's blood supply. In addition, researchers have been exploring the role of cholesterol in Alzheimer's disease for years because it regulates both the generation and clearance of beta-amyloid, the main ingredient in Alzheimer's plaques. Doctors also thought Jerry might explore the use of tacrine, a drug that had recently been approved by the FDA for mild cognitive impairment.

The doctors recommended against Jerry pursuing his pilot's license, which they deemed too dangerous; in fact, they thought he probably shouldn't even drive a car much anymore.

There would be no more airplane. No more tinkering in the garage. No more teaching technical college. At forty-nine, in one short medical visit, he'd gone from being the man of the house to a dependent. Soon, he wouldn't even be trusted to drive the cars he once knew how to build. Yet, devastating as this was, Jerry managed to take it in stride. He had suspected something was wrong, but he intended now to make plans that would cushion his family as he declined.

"In some ways, it was kind of a relief to him to know," his wife said. "I think he didn't feel like he had to hide things and pretend everything was all right."

Jerry and Sharon had always been practical, and they jumped into action as soon as they got home. He took early retirement from OSU Tech; they arranged for retirement and disability payments. They also purchased a piece of property in Beggs, Oklahoma, that was large enough to create the family compound Jerry had long dreamed of. In earlier, happier times, they had imagined they would grow old in the country, surrounded by the laughter of their grandchildren. But now there were practical reasons for the move: More family members would be nearby to help as Jerry required more care.

"He never showed to me how hurt or worried he was," Sheryl recalled. Her father remained upbeat, as he always had been. Just as he had once fed the needy, he thought of his disease as an opportunity to help other people; he wanted to participate in research to help find a cure.

The drug the NIMH recommended for Jerry, tacrine, was the first ever approved by the FDA that had been developed specifically to treat

Alzheimer's symptoms. Sold under the brand name Cognex, it was a cholinesterase inhibitor, which prevents the breakdown of a chemical called acetylcholine, which aids memory, thinking, and reasoning.

Cognex followed a circuitous route before it got to Jerry DeMoe, since federal regulators did not exactly welcome the drug with open arms. In 1991, an FDA advisory committee rejected Cognex, saying its side effects—including liver damage—outweighed any small benefit it might offer. Although its inventor, a California doctor named William K. Summers, had claimed that Cognex dramatically helped Alzheimer's patients and downplayed the side effects, the FDA cast doubt on his research methods and conclusions.

For the Warner-Lambert company, which had hoped to market Cognex, this setback was devastating. At least one investment analyst estimated that the drug would have been worth $1 billion in annual sales.

But the disappointment didn't last long. Two years later, the FDA reversed itself, and the advisory board said despite the side effects and questionable benefits, the drug should be available to patients because they basically had no other options. At least one panel member called it "a matter of conscience."

Summers, who took a sabbatical to promote the drug's approval to the FDA, hailed the reversal as a vindication, calling it "a great day for Alzheimer's patients and their families." He predicted that once ordinary doctors got their hands on Cognex, the world would see the amazing benefits he had reported in his studies. Jerry DeMoe began taking the drug in 1995.

But Jerry never saw the improvements that Summers promised. Plaques and tangles continued to collect in his brain, and his symptoms continued to worsen.

Soon after the diagnosis, Sharon made herself a vow: She would care for Jerry at home, and would never send him away, as Gail had done with Moe. When they moved to the new property in Beggs, Sheryl, who was now fourteen, would start at a public high school to free up Sharon's time for her husband. It was a scary transition for Sheryl, a shy child who wasn't sure what to expect from her new home or her new school. And she was saddened to leave their old house, where she had spent every holiday, birthday, and family gathering of her young life. She described it as "a pretty and sweet season of my life, gone."

Jerry was still well enough to drive his daughter to school, but his speech was becoming difficult to understand; it was one of the first major pieces of his character to disappear.

As she watched Jerry working on his old GMC truck, tinkering with his tools, Sheryl saw his frustration. A simple sentence—*I'm going outside*—would turn into "a long, exhausting guessing game," she said. He compensated by poking fun at himself when he forgot a word, trying to laugh away the embarrassment, the way his North Dakota family might have. Though she had told herself she would stick by him, Sheryl grew impatient and began spending less time with her dad and more with the new school friends she eventually made.

"I knew this was one of the stages of the disease, but I hated it," she would later write. "This wasn't my dad, he was perfect, and this man can't even talk normal. What happened?"

As the difficulties of her own teenage years took hold, she began withdrawing from both of her parents, often acting selfish and rude.

She hoped for the impossible, that one day her father would go to the doctor's office and learn that the drug he was taking had stopped his disease. Every night she prayed: *Please make him better.*

It never happened. When their church held a laying-of-the-hands service to heal him, Jerry had his only moment of clarity in what felt like years, and became very emotional. Sharon was deeply moved by the service. But Sheryl, a skeptical adolescent, was frustrated that God had not yet cured him.

Outwardly, her parents maintained their composure. Her mother asked her to talk about her feelings, but Sheryl shut her out. What she didn't know was that privately, Jerry's vast reserves of positive thinking were rapidly being depleted; to his wife, he talked about suicide. Behind her composed façade, Sharon was thinking about it, too. She turned to her faith to carry her through. When she spoke to God, her request was simple.

"I mostly wanted help; how to help him, and how to help the situation," she said. "My prayers became shorter, and I was desperate."

Gail called to offer her support. To Sharon, it was a lifeline, even when they didn't talk about the disease. They laughed and joked; Gail believed in using humor to relieve some of the emotional burden.

"We knew the pain. I knew what she felt, and she knew what I felt," Sharon said. "It was someone who I knew understood completely."

As Jerry's disease progressed, he became irritable and angry, as had his brother before him. He spent a lot of time in front of the television. Jerry's grandchildren adored him; they called him "J-Pa" (Sharon was "Meemaw"). Though he had once loved children, their noise and commotion began to irritate him.

At the same time, Sheryl was consumed with shame over how she was treating her parents, but she couldn't seem to break free of it; every day, her father only got worse. The Cognex had not slowed down the Alzheimer's progression, and it was apparent that he was fighting a losing battle. He tried to fling open the car door when his wife was driving; he ran away from church. Sometimes, he hid outside their house where Sharon could not find him.

Miserable at home, Sheryl found solace in the company of her boyfriend. In January of their senior year in high school, he proposed; he even asked Jerry and Sharon's permission to marry their younger daughter. Jerry was lucid enough to give his blessing. Like her cousin Karla before her, Sheryl married the man she had been dating since eighth grade.

In August 2000, a radiant eighteen-year-old Sheryl wore a sleeveless white gown, one hand carrying a bouquet of yellow roses, the other hand tucked in the crook of her proud father's arm. Lori McIntyre made the trip to see her young cousin's wedding. But when the preacher asked Jerry the traditional question—"Who is giving this woman in marriage?"—he could not find the words to answer. His wife said them on his behalf, then helped him sit down.

Before the couple left on their honeymoon the next day, Jerry pulled his new son-in-law aside to say, "Take care of my little girl." A dozen years and four babies later, the thought of it still makes Sheryl cry.

• • •

After Sheryl moved out, Jerry became more aggressive, more bewildered. As had been the case so many years ago with his uncles—Wanda DeMoe's brothers, who questioned their own images in the mirror—he became angry at his reflection, so Sharon covered all the mirrors in the house with sheets. It felt as if they were mourning him already, a Jewish family sitting shiva for the dead. He confused his wife with Wanda, his late mother. Sometimes, angry and belligerent, he grabbed her, but Sharon was not afraid. It was not until he failed to recognize her as anyone but a stranger that she finally broke down.

She wished she could detach herself from her husband enough to send him to a nursing home, the way Gail had, a decision she now understood better. And while many people described Gail as one of the strongest people they had ever met, Sharon wasn't so sure.

"I know how that goes. You can really make people believe that," she said. "I know there were weaknesses, but she hid them well."

Although Jerry comprehended less and less as time went on, he was aware enough of his surroundings to feel the sting when the neurologist in Tulsa who had been running his tests for the research trials determined that his contributions were no longer useful; he had deteriorated to the point where he couldn't answer questions. He was devastated by the idea that he could no longer help.

"He really wanted to do something," Sharon said. "In his mind, that was very important, and I know it is to all of the DeMoes. That was a setback."

The problem came to a head when, at one appointment, the doctor entered with an unfamiliar colleague and startled Jerry; Sharon slid in front of him, but it was too late to prevent Jerry from getting agitated and upset. The doctor, who was worried about his own safety, called security and suggested a psychiatric ward.

Sharon froze, not knowing what to do, and they took Jerry away and dosed him with Haldol, a powerful antipsychotic. She had no idea what her rights were, or how to advocate for him.

"They completely knocked him out," said Sharon. "He didn't even do anything. I mean, I could have handled it. I don't know what the doctor thought. They wouldn't let me get him out."

She went to visit him, a newly pregnant Sheryl in tow; he asked to be taken home, but the hospital hadn't released him yet. To distract him, Sharon turned his attention to their daughter.

"Sheryl's here. Isn't she getting fat?" Sharon teased.

Not understanding the joke, Jerry gave his wife a dirty look. But Sheryl's laugh relaxed him, and they explained.

She was having a boy, Sharon said. Jerry was finally going to get his boy.

When Jerry was released, instead of placing him in an institution, Sharon brought him home and found another doctor. That arrangement worked briefly, but the ugly truth about Alzheimer's, of course, is that it never gets better. It can be contained and controlled for a while, sometimes for months, sometimes for years; but in the end, the disease always wins, and the symptoms boil over, burning whoever is attending the pot. Sharon would learn, as many caregivers do, that it's not a question of love or devotion or commitment. Finally, she caved and called a nursing home.

Her intent was to have him stay there for two weeks, just enough time for her to rest. She was exhausted. But then he tried to escape and was moved to a different facility. He asked to come home.

By the time Sheryl and her husband welcomed their son on June 21, 2002, and brought him to meet his grandfather, Jerry didn't even know they were there.

After a brief hospital stay, Sharon contacted a hospice, installed a hospital bed in their living room, and brought him home.

"I thought he was going to live for years like that, and I was willing to take care of him that way," Sharon said. "I don't know if I was in denial. Nobody really sat me down and told me that he was really bad . . . I thought I'd rather have him sick than not at all."

In total, Jerry DeMoe lasted at home for only about ten days. Years before he ever developed Alzheimer's disease, he and Sharon had discussed their end-of-life wishes; neither of them wanted life support or to be resuscitated when their time came.

One night, Sharon pulled her air mattress beside the bars of his hospital bed.

"I wanted to get in bed with him, just lay beside him," she said. "I climbed in there. Well, I couldn't get out. I weighed more than he did at the time." She laughed fondly at the memory. She had eventually struggled her way out. "But I got to cuddle him a little, anyway," she said.

Thinking about this man she had loved her entire adult life, she still felt the unspoken magic that had brought and kept them together.

"He was a sweet guy; he was caring. He was macho and masculine, but yet he had a soft heart and cared about people," she mused. "I was a spoiled brat, actually. I don't know how he put up with me, because he wasn't that way.

"I guess he loved me."

On September 17, 2003, Sharon called Sheryl to say: Come home. It's time to say good-bye.

Numb, Sheryl was unsure what to expect. It wasn't like the movies; there were no comforting final words, no peaceful deep sleep. There was only a shell of her father, once so vibrant and loving. She wanted to tell him how sorry she was for how she had acted toward him for all those years.

At some point, in those final hours, she did hold his hand. It was cold and skinny, not the warm, strong grip of the man who'd held her as a little girl. She could not make herself speak; she hoped that he felt her love through that hand.

Nine

THE FRUITS OF PERSISTENCE

BY 1995, HAVING isolated three autosomal dominant mutations that guaranteed Alzheimer's disease—APP, PS1, and PS2—science had a small group of patients who could potentially offer some clues about the way the disease developed. With a simple blood test, doctors could identify that a person had one of the mutations and was definitely going to get the disease.

In the years that followed, time became the chief roadblock to finding a prevention therapy for Alzheimer's. The standard approach would be to gather thousands of volunteers, including healthy people; give them experimental treatments and placebos; then see what drugs budged symptoms in people who ultimately developed dementia. But that process required too many healthy volunteers, too much money, and too many years—longer than the life of most drug companies' patents—to achieve any meaningful results.

Such trials were attempted. The National Institutes of Health conducted its Alzheimer's Disease Anti-inflammatory Prevention Trial, nicknamed ADAPT, on twenty-four hundred elderly volunteers beginning in 2001.

Participants were given anti-inflammatory drugs, and the study's research team tracked them to see whether the medications cut down on the risk of Alzheimer's after age seventy. Concerns over cardiovascular side effects sidelined that study in 2004, but the participants were followed for seven more years before researchers concluded that the drugs didn't help them. Meanwhile, as a generation of baby boomers continued to age, a tidal wave was building. Science couldn't afford to wait.

In addition to the logistical problems facing the field, a debate began brewing among scientists over the key mechanisms responsible for the disease. On one side, which became the prevailing view, researchers believed that an excess of beta-amyloid protein caused the disease. This theory gained traction after UC San Diego pathologist George Glenner detailed the structure of beta-amyloid in 1984. Since Alzheimer's patients had plaques forming outside their brain cells and beta-amyloid was the plaques' main ingredient, a hypothesis formed that beta-amyloid played an important part in the disease's development. The fact that genetic mutations led to overproduction of beta-amyloid seemed to support this theory, and in time, scientists who adhered to it were nicknamed the "Baptists" (for the first initials of beta-amyloid protein).

But the other side argued that neurofibrillary tangles forming inside the brain cells, and not plaques, were the primary culprit. This group was called the "tauists," after the tau protein, the chief ingredient in the tangles.

In a healthy brain, cells receive nutrients through a transport system made up of structures called microtubules, which run in parallel lines that look like railroad tracks. The tau protein, which normally has some phosphate molecules attached to it, binds to the microtubules and stabilizes them to keep everything running smoothly.

In an Alzheimer's brain, an abnormally high amount of phosphate molecules attach to the tau, causing it to break off from the microtubules and attach to other tau threads, forming tangles inside the cell. Without tau, the microtubules disintegrate, and the cell becomes unable to communicate with other neurons. Eventually, the neuron dies. With it die the connections networking it to other neurons and the synapses that transport memories. When enough cells die, the affected parts of the brain begin to shrink.

Regardless of whether amyloid or tau was the primary disease trigger, researchers also faced another major obstacle: Nobody had any way to look at Alzheimer's in a living brain. They had to wait until the patient died, then examine brain tissue during an autopsy.

That made perfecting any kind of drug treatment maddeningly difficult. In order to stop the disease, scientists needed to track the way it developed so they could try to predict its path and test treatments. Without being able to see the brain's response to the drugs, neurologists wouldn't be able to pinpoint what was and wasn't working. If they could figure out a way to fix their brain-imaging problem, researchers would be able to watch the disease unfold in real time, which would help them narrow their approaches for drug therapy.

But basic biology was working against them. The body's natural filtering system, known as the blood-brain barrier, prevents most substances (including anything with an electron charge) from crossing into the brain through a series of tightly joined cells.

In 1994, this was the challenge facing Chet Mathis, a radiochemist at the University of Pittsburgh who specialized in developing compounds to image the brain. He'd arrived there two years earlier because the university had a well-respected medical school stocked with psychiatrists who were itching for someone to help them create images of serotonin and other neuroreceptors, which are the chemicals the body produces to communicate information throughout itself. Today Mathis speaks reverently of the school's dedicated cyclotron—a one-hundred-ton, $2 million particle accelerator used in physics—the way other professors speak of perks like parking spaces and posh faculty clubs.

Mathis was passionately in love with his work, so much so that as he warmed to his topic, he tended to speak very rapidly. His sense of humor was wry, and he had permanent crow's-feet that deepened when he laughed. It was an honest laugh, the kind you had to earn.

His office at the University of Pittsburgh Medical Center's sprawling Presbyterian hospital was buried deep on what he called the ninth-and-a-half floor, in a suite where the laboratory was lined with lead and copper to keep radiation in check. His office wall had been knocked down to allow cranes to hoist in a new cyclotron, and his office window looked out across the street to the yellow brick façade of the Western Psychiatric Institute and Clinic, where Bill Klunk, one of his collaborators, worked.

Mathis had begun his career spending much of the early eighties helping to make fluorodeoxyglucose. FDG is a radioactive chemical used in imaging to diagnose, among other things, lung cancer, breast cancer, and melanoma. In those early years, scientists were also starting to try it out as a diagnostic tool for Alzheimer's patients.

Eventually, at a neurologist's request, Mathis began to radiolabel a dye called Congo Red that was tested in dead patients. By adding radioactive atoms to its chemical composition, scientists can track a dye as it works through the body. It can then, theoretically, be manipulated further by altering its chemical makeup so it will stick to certain proteins—for example, amyloid beta or tau.

In the early part of his career and for many years after, Mathis continued working with Congo Red. As its name suggests, it's a bright red, toxic sodium salt first created in 1883 by a German scientist who was trying to create a textile dye. The company didn't like its gory color, so the scientist sold it to the AGFA company of Berlin, which christened the dye "Congo Red" in honor of the 1884 Berlin West Africa Conference, a key event in the colonization of Africa.

One hundred or so years later, Congo Red found a new purpose as a dye used in radiology for imaging. Scientists discovered that its chemical properties allowed it to stick to amyloids in much the same way it once bound to textiles.

The problem for Mathis, though, was that it refused to cross the blood-brain barrier, making it pretty much worthless for a study of amyloid in the brain. Finally, he gave up on the dye, but the failure grated on him for years.

One day, however, a young psychiatry professor named Bill Klunk walked into Mathis's office at Pitt, saying, "I've got a new compound. It's much better than Congo Red."

Klunk's journey had begun in 1987. A shy, soft-spoken native of Hanover, Pennsylvania, he was fresh out of Washington University in St. Louis, having been recruited by David Kupfer, who was then the chairman of the University of Pittsburgh's Department of Psychiatry.

While he was still at Washington University, Klunk had become interested in the biochemistry of the brain and how it changes in people with dementia. Like Mathis, he started with the erratic Congo Red, which researchers thought might keep beta-amyloid protein molecules from sticking together and forming plaques. It didn't.

But he kept monkeying with the chemistry of the compound, hoping that one of its derivatives would at least help him take better pictures of the brains of Alzheimer's patients.

"I had small grants here and there, and I kept this interest active at a low level until we got to the point of writing a paper about what we had done," recalled Klunk.

That was in 1994, and Klunk was thinking about patenting some of the technologies used in his research to date. As he talked to Pitt officials about his idea, they referred him to Mathis. It was that series of Congo Red derivatives that he pitched when they met.

Mathis agreed to work with him, and they started with a few grants from the Alzheimer's Association. Eventually, a drug company became interested and provided more funding. But they weren't exactly working at a breakneck pace.

"We had years of just sort of crawling towards our goal, from '94 to '99, in this Congo Red series of compounds," said Klunk, who found that he wasn't having any greater success with the dye than Mathis had. "There were a lot of lean years here. We made hundreds of compounds that failed."

Those failures—more than three hundred, in fact—didn't just bring frustration; they also carried a certain amount of risk to Klunk's career. In academia, when a scientist is up for tenure and promotion, the university typically asks people in related fields to provide feedback based on the candidate's track record of publications and grants; hence the maxim "publish or perish."

For most of the century, science had been unable to view Alzheimer's inside the brain, matching the internal progression of plaques and tangles with the deterioration of a person's memory and function. The field was desperate for a closer look at the disease it was fighting, hoping to gain a more precise measurement of how effective different therapies were. Unfortunately, too many of them were afraid that the quest would kill their careers in the process.

But what if the scientist, instead of worrying about establishing an elaborate track record, decided to pursue the holy grail? What if he decided that he was not going to make a superficial stab at it; he was going to devote all his time to a single huge feat, trying and failing over and over again? And what if his efforts, though noble, didn't yield many papers or grants?

In some academic settings, the university would have pulled the plug. But Klunk was lucky: Kupfer had the faith to stand by him, and a handful of colleagues in the field quietly pleaded with the university, in their independent evaluations, to allow him to keep going, knowing how important his work might be to the rest of them.

The decision to finally abandon Congo Red altogether came either from frustration or altruism, depending on whom you ask. The drug company that had funded Klunk and Mathis's research started losing interest. It was time to try another tack.

When Klunk had started in 1987, he ordered the three basic compounds that are known to work with amyloids: Congo Red, Thioflavin T, and Thioflavin S. As he turned to his remaining two options, he asked Mathis a question: What properties do you need for imaging?

Mathis's number one priority: It had to get into the living brain.

Most dyes like Congo Red carry an electron charge, which is what makes them bind to things like textile fibers, the purpose for which they were originally invented. To get them past the barrier, scientists must experiment with attaching them to molecules that can cross over, such as glucose or amino acids. But these efforts are only intermittently successful. The difference between success and failure sometimes rests on a single atom.

Mathis's ideal dye couldn't just leak into the brain in trace amounts; enough had to get through to give a full and accurate image in a brain scan. And it had to be nimble, too, showing up in the brain within two to three minutes of being injected into the patient's arm, then quickly exiting the body without damaging anything in the process.

As they moved back to the drawing board in 1999, Klunk and Mathis started with Thioflavin T, which was a simpler molecule and easier to make since it only had one electron charge.

Finally, after years of crawling, they broke through with that one change: In fact, they were on their feet and sprinting. Five months after switching to Thioflavin T, they found entry into the brain.

"Whether it was blind luck or the fruits of persistence, no one can say," Mathis and Klunk wrote in a paper about the discovery. It didn't matter. Within a year, they had a hundred different derivatives, including one that would come to be known as Pittsburgh Compound B, or PiB.

When they injected the new compound into mice, it worked at a rate ten times better than any of their previous efforts. They already knew, through trial and error, that it would stick to the beta-amyloid proteins—although they still aren't sure why. They don't care, either, because when it gets into the Alzheimer's brain, it illuminates plaques on a scan in glowing red, a brilliantly displayed road map of the disease.

After testing the compound's ability to highlight amyloid on human cadavers, they were ready for clinical trials in living patients. Klunk and Mathis contacted Bengt Långström, the director of the Uppsala University PET Center in Sweden, who was leading the field of positron emission technology (PET) imaging with microdoses, or less than ten milligrams, of a radiotracer—the equivalent of about one-ten-thousandth the amount of drug found in a low-dose aspirin.

The study was scheduled for Valentine's Day 2002. Klunk and Mathis would never know the name of the first person dosed with PiB; she was only identified as "AD2." They knew simply that she was relatively young and that her memory problems had forced her to give up her career as a health care professional. Her cognitive testing scores were just below normal. And she knowingly agreed to be their guinea pig.

Throughout that day, Klunk and Mathis waited on tenterhooks in Pittsburgh for any word from Sweden, which was six hours ahead. And then Långström phoned.

"I think we have something to celebrate," he said.

Through the Swedish research team, Klunk and Mathis wrote a letter to that first, anonymous volunteer:

It is only through selfless individuals such as you that medical scientists can push ahead the frontiers of research, they wrote. *You should take pride in knowing that you played an important role in allowing these changes to begin. Thank you very much.*

The patient known only as AD2 died five years later, in late 2007, from Alzheimer's disease.

The compound worked just as well in the brains of other Alzheimer's patients. Klunk watched, breathless, as PiB left the bloodstream, crossed into the brain, found plaques, and then worked its way into the center of the plaque. The entire process took about ten seconds.

The only thing that could hold them back now was if a company became interested in their work and argued over intellectual property rights to the new family of compounds. That hadn't happened yet, but now that they were finally having some success, Mathis didn't want to risk it. He had finally conquered the problem that had been dogging him for most of his professional life, and now, just when his research was starting to accelerate, a relatively minor scuffle over intellectual property could throw down spike strips.

He told Klunk he wanted to publish their results and make them part of the public domain, effectively voiding any company's interest in patenting the compound. End of problem.

But Klunk was more pragmatic. "Yeah, we could do that," he told Mathis. "But the ultimate goal is to help people, and if we did that, no company would pick it up."

Mathis hadn't thought about it, but it was true. Once their research was publicly disclosed, for all intents and purposes, it would have no commercial value. The reality of drug research in the United States is that without corporate funding to develop it, PiB would never find its way to the millions of Alzheimer's patients waiting for a diagnosis. And nothing that followed—the potential for trying out preventive drugs—would happen, either.

"It shows you: He was thinking about that," Mathis said of his partner, with pride in his voice.

Since they were university employees, Pitt had the right of first refusal for their work. Under the terms of the agreement, Mathis and Klunk were supposed to be paid one dollar. (They never got it.) Pitt then sold it to a British company called Amersham that was acquired by General Electric in 2004.

Klunk resisted praise for the decision. Of course he was thinking about the patients, he pointed out: He was a psychiatrist. Mathis was a radiochemist, so he was thinking about the science. And that, he said, was the key to their partnership.

Meanwhile, the research field exploded. The discovery of PiB "represented a major breakthrough," recalled Kupfer, the department chair who had stood by Klunk through a decade of failure. "It offered a new sense of hope—not only to patients and families, but to the whole investigation field."

Moreover, the breakthrough justified Kupfer's faith in the two men, proving that risky projects could still carry a huge payoff.

GE Healthcare licensed PiB with plans to bring it to commercial market. Meanwhile, dozens of other research institutions around the world used it, and in 2012, the FDA approved Amyvid, a PiB cousin with a longer half-life that belongs to the pharmaceutical giant Lilly.

Mathis and Klunk continued to work on other compounds, including one funded by actor Michael J. Fox to take images of Parkinson's disease. They were also trying to image tau, the protein that makes up tangles, since PiB focuses only on plaques.

In 2008, for their discovery of PiB, Klunk and Mathis were awarded the American Academy of Neurology's Potamkin Prize, which is nicknamed "the Nobel Prize of neurology" and recognizes achievements in research for Pick's, Alzheimer's, and related diseases.

"Bill always says we got really lucky," said Mathis. "I say, 'No, Bill, we didn't get lucky. We were due success.'"

Pondering that thought, Klunk finally agreed. To a scientist, serial disappointment is the price of admission for a moment of great discovery.

In other words, to triumph, "you've got to fail," said Klunk.

Ten

DÉJÀ VU

IN NOVEMBER 2003, Karla watched a PBS documentary called *The Forgetting: A Portrait of Alzheimer's*. At the small college where she worked as an administrative assistant, a psychology professor gave her the film, knowing her father had died from Alzheimer's. *The Forgetting* featured Bill Klunk, who talked about his discovery of Pittsburgh Compound B and what it could mean for early diagnosis and prevention.

The Forgetting also introduced a family from Massachusetts, the Noonans. They were ten siblings who were roughly of an age with the DeMoe kids. Their mother had developed strange symptoms starting when she was thirty-nine; doctors thought it might be a nervous breakdown or some acute form of depression. Like Wanda DeMoe, she was given shock treatments. Ultimately, she died of pneumonia.

One of her daughters, Julie Noonan Lawson, was interviewed in the documentary. Like Karla, she wore her blond hair in a short bob. After their mother died, the children thought they had closed that tragic chapter of their lives, Julie explained—until decades later, when her older sisters, Fran and Maureen, also started showing symptoms.

"That's when we realized: We're not done. That's when I realized the magnitude of this disease," Julie told the camera. "This is gonna hit us again."

The parallels were astonishing. Karla was riveted; she watched *The Forgetting* over and over and showed it to her mother. The documentary explained that the Noonans' affliction was caused by a rare genetic mutation. Was her family like the Noonans? Was this what the doctor meant when he told her mother that half of the children might get it? Had some of them inherited this disease that had taken their father?

In February 2004, Gail received a disturbingly familiar phone call from a coworker of Brian and Doug.

"Gail, I think there's something wrong with your boys," he confessed. "They aren't able to do their jobs like they were."

The news didn't come as a complete surprise; for a few months, the family had been making the same observation. Since his late twenties, Doug—who, ironically, shared a birthday with his uncle Jerry—had been repeating himself. And while everybody in town noticed it, nobody wanted to believe the disease was coming back for another round. But fear was beginning to simmer in their collective subconscious.

Gail and Brian visited Karla in Fargo for Christmas just two months before the phone call. As they had done eight years earlier, when they saw Jerry at the family reunion, the women discussed some of the changes they'd observed in Doug. Despite her earlier efforts at denial, and perhaps motivated by what she'd seen in *The Forgetting*, Karla thought it would be a good idea to get a doctor's opinion.

Brian, who was sitting with them, didn't say anything. "When I think about it now, he was there, but he wasn't really in the conversation," Karla said. "We knew Brian wasn't like he should be, but he had smoked pot for so many years. So we thought the pot did it . . . I never, ever in my mind would have thought Brian had the disease."

For fifteen years, ever since Moe's death, Karla had been living in a state of denial. Even after Jerry's diagnosis, she refused to think about her generation dealing with Alzheimer's. She banished the thought any time it tried to enter her brain, as though she could will the possibility into submission. It wasn't fair, after what they'd already endured. Their happy childhood had been cut short; hadn't they at least earned the right to a happy ending?

But that February phone call changed everything.

Gail and Karla enlisted Dean to design an elaborate ruse that would get the boys to submit to testing. Karla found a neurologist's office in Fargo and described her situation to the staff; miraculously, they offered an appointment in just a few weeks. Brian and Doug would come in for three days of interviews, neurological testing, and scans at a hospital, believing—with the doctors' cooperation—that the entire family would undergo the same tests because of their history. Dean, who was now living on Karla's side of the state, near Grand Forks, reinforced the cover story by taking time off from work to attend. In reality, he was only there for moral support.

Ever since Dean's move, he and Karla had grown closer, just as he had visited Lori often when he was living in Colorado. Both Dean and Karla were deeply involved parents, and their kids were near enough in age to relate to one another. If bad news emerged from the testing, they vowed to team up and become their brothers' caretakers. As Gail grew older, it only made sense for someone from their generation to assume the burdens she had once shouldered for their father, and neither Brian nor Doug had wives who might look after them. Lori and Steve were still moving according to the whims of his railroad job, but Karla and Dean were now firmly planted in North Dakota and raising their kids there.

They were also both organized, determined, capable by nature. And it was easier for Karla to brace herself for the possibility of bad news knowing that she had Dean to depend on for help.

As the DeMoes sat in the neurologist's office in March 2004, the doctor did his best to reassure the family. No matter what, he told them, everyone would feel good about themselves by the time they left. It would be a relief to know where they stood genetically.

But on the last day of testing, the neurologist was out of town, so another member of his team—a neuropsychologist—sat the family down to discuss Doug's and Brian's results, with the two men in the room. His comments were stunningly insensitive.

"They have the math acumen of a third or fourth grader," the psychologist said, speaking to the rest of the group as if Brian and Doug weren't there, or were perhaps too stupid to understand. He displayed slides of the childish scrawls the two men had made in their cognitive test answers. He pointed out how poorly they'd performed: "Brian has the reading ability of a third grader and the spelling ability of a second grader." Brian was nearly forty-eight; Doug was forty-three.

It was embarrassing, having the men of the family depicted like slow children. The reality of their situation was harsh enough, but the doctor seemed unaware of how much worse it felt to be told such devastating news as though Brian and Doug were nothing more than lab specimens.

The imaging technique developed by Bill Klunk and Chet Mathis was not yet available, so the diagnosis was based on a clinical assessment: It was Alzheimer's disease—the same thing that had stricken Moe, his brother Jerry, their mother, and their uncles.

Gail began sobbing; then Karla also broke down. Dean and his wife, Deb, were incensed at the doctor's detachment. Doug and Brian just sat, listening; Karla thought perhaps they didn't comprehend what had been said about them.

The psychologist asked if he could draw their blood and test them for a possible genetic mutation, to see if that was the root cause of the disease in their family. *Like the Noonans*, Karla thought.

The family agreed, then asked to be referred to any Alzheimer's research studies that might use them as volunteers. When tragic news breaks, people respond in many ways. In the DeMoes' case, their immediate response was: Fight back.

On their way out to the car, Brian looked at Doug and said, "Boy, are we *dumb*, Doug."

They had understood.

• • •

Setting aside their grief over the diagnosis and their anger over the way they were told, the family did take away an important kernel of information: Just as Karla had suspected, their problem might come from the same type of rare genetic mutation that was in the documentary. A month later, the neuropsychologist contacted Karla, confirming that Doug and Brian had tested positive for just such a mutation.

Doug's daughter, Jennifer, was barely out of high school and living ninety minutes away in Minot when Doug Latch, the father she so adored, had started to falter. When she talked to him on the phone, they'd get stuck in an endless conversational loop; he'd tell her five times in a row that he had washed his car.

She hadn't known her grandpa Moe very well, since he died when she was just a little girl. Nobody ever really talked about him, except to say he was a bearcat. When Doug was diagnosed, she was shocked.

"I don't think anybody thought it would turn into this. I thought only old people got this," she said. She remembered thinking: "Holy crap. This is going to turn our world upside down. Nothing's ever going to be the same again."

She moved back to Tioga and into Doug's tidy little house, with no specific plan for what she would do next.

"I just thought I had to be here," she said. "There's just not a lot of years that we were ever together."

Though the disease took Jennifer by surprise, one of her cousins—Lori and Steve's oldest daughter, Jessica—had the opposite reaction. Jessica was now in her early twenties and living in Denver. When she first learned that her uncles carried the gene, she went numb, the natural pessimist in her taking over.

"I thought, 'Oh crap, *both* of them?' And I thought in my mind that was going to happen to everybody," she said.

Based in the northeast, the Noonans had been able to turn for help to Massachusetts General Hospital; in North Dakota, the DeMoes had struggled to simply find a neurologist with the genetic expertise they needed.

Tioga didn't even have a McDonald's, let alone an Alzheimer's research facility.

"I knew we needed to go someplace bigger to see if we could get help," Karla said. Remembering the study to which her father had contributed through the University of Minnesota, she tried to locate June White, the nurse who had drawn their blood, not realizing that White had retired in 1995. Every road turned out to be a dead end. From the DeMoes' perspective, it was as though that research had never occurred.

Vic, Moe's younger brother who was closest to him in age, recalled giving blood, too. What he didn't remember was getting any meaningful information from the study, and it pissed him off. Somebody transferred jobs somewhere, he thought, and nothing ever came of that early hope. So while he still had his memory tested every year at Indiana University, he refused to donate money to the Alzheimer's Association.

"I've been doing this for fifty years, and you're no farther ahead than you were!" he told them. "These people. When they run out of grant money, they quit." And the test subjects who had given of themselves, many of them desperate, were left stranded.

Like other Alzheimer's patients and family members who contribute to research, the DeMoes were finding that they could often be left at loose ends by scientists who ran out of funds or moved on to other jobs. The patients contributed their tissues, their blood, even their brains after death. They revealed some of the most intimate details about their daily struggles, their fears, their family histories. And then, if a study was orphaned, it seemed as though all that information fell into a black hole, the contribution wasted.

Karla DeMoe Hornstein had been many things in her life: a daddy's girl, a self-centered sister, a homecoming queen. But when her family needed her, she reached underneath the security of her cheerful façade, elbow-deep into her soul, and found at her core the strength that Gail had been instilling gradually in her for her entire life. When she pulled that strength to the surface, she was permanently altered. Gone was the woman who deferred to others' opinions and accepted no for an answer. She started searching online for Alzheimer's research and came across the National Institute of Mental

Health, which she knew was where her uncle Jerry had gone when he was diagnosed. She started making phone calls.

And finally—finally—someone called her back. It was a woman who said: "I knew your uncle. I remember him. We'd like to meet you."

In August 2004, five months after visiting the doctor in Fargo, Doug and Brian went to Maryland with Karla and Dean to undergo a more detailed examination of their brains than they ever previously had thought possible.

They were introduced to Dr. Pearson Sunderland III, chief of geriatric psychiatry and a longtime Alzheimer's researcher. Known professionally and personally as Trey, he was handsome and charismatic, and he treated them well, making them feel as though they were in competent hands.

Sunderland had arrived at the National Institute of Mental Health, a subset of the National Institutes of Health, in 1982 for a fellowship, then begun a research program in Alzheimer's disease and geriatric depression. He spent the next twenty years building a sterling reputation, including a decade as the chair of the NIMH's institutional review board and head of the medical advisory board for the Washington, DC, Alzheimer's Association. He was a prolific writer, authoring more than 250 scientific papers as well as a book on Alzheimer's. It seemed the DeMoes had found their expert.

Sunderland's specialty was finding ways to detect Alzheimer's before patients developed symptoms. As with most formidable diseases, finding a treatment for Alzheimer's seemed most attainable in its earliest stages. However, no drugs to date had been able to halt its progress. Once it began destroying neurons, its effects were irreversible—doctors called it "the cascade." The key, then, seemed to be preventing the disease before it reached that fatal tipping point.

But the Alzheimer's brain changes years before it's apparent to the outside world. Nobody even knew what presymptomatic Alzheimer's looked like, or whom to study, until they found people like the DeMoes, who carried a genetic flaw that guaranteed a future disease. Now, in addition to that population, they had Pittsburgh Compound B, a way to see what the disease looked like before it was evident in a person's behavior. The field planned to combine the two to refine a target for possible drug interventions.

Beginning in the early 1980s, Sunderland began studying Alzheimer's in patients who came to the NIH Clinical Center, the world's largest medical experiment facility. He and his staff collected their spinal fluid in search of biomarkers, or biologically distinct indicators of the disease, such as beta-amyloid and tau proteins. By the time he met the DeMoes, he had collected more than six hundred samples.

In the four days they were in Maryland, the DeMoe siblings learned many important but alarming things about the disease that had been stalking their family for generations. The fact that it was autosomal dominant meant that any of them could have inherited the mutation from their father. If they did, Alzheimer's disease was guaranteed, and its symptoms would begin showing soon; all of them except Jamie were now in their forties, which was close to the expected age of onset.

That genetic guarantee was a double-edged sword. For science, it meant researchers had perfect specimens to study. But for the DeMoes, it was a catastrophe.

Suddenly, the notion that they all had decades left to spend raising their children, making house payments, and earning paychecks vanished. Worse still, if they had it, there was a fifty-fifty chance they were going to pass it on to their own children, or already had. And, except for Jamie, all of them had kids, much-loved children born into a world with the automatic expectation of good health and long lives. Now those children were suddenly in danger: Kids who were still playing Little League and toting around stuffed animals might be unable to have children without the very serious risk of passing on a fatal illness. The grandchildren their parents dreamed about would be equally at risk.

To be able to predict one's own death is a weighty business; it affects decisions both before and after the discovery, and its ripple effect touches parents, children, siblings, and family members. It is knowledge that can affect the most profound life choices: whether to marry, have children, participate in research, take a once-in-a-lifetime trip that you can't really afford. It also

Ruth "Wanda" Hubbard, Moe's mother (bottom row, second from left), graduating high school in 1924. DeMoe family

Galen "Moe" DeMoe, left, and his brother Vic. Moe would inherit the disease that plagued their mother; Vic did not. DeMoe family

Galen DeMoe's high-school graduation photo, Eau Claire, Wisconsin. He wrote frequently to Gail, whom he met while working in the oil fields of North Dakota during the summer. DeMoe family

Gail Helming's high-school graduation photo, 1953. She was the valedictorian of her class of six, and her romance with Moe was well underway. DeMoe family

Gail, seated left, visited Moe's family in Wisconsin. Wanda DeMoe, seated right, acted strangely—but Gail dismissed the odd behavior because she was so in love. Standing behind her is Moe's young sister, Pat. DEMOE FAMILY

Gail and Moe's wedding shivaree, August 27, 1955. DEMOE FAMILY

Moe and his oldest child, Brian. Gail would always remember how much Moe loved kids, but their own children grew up to remember him much differently. DEMOE FAMILY

Doug, in an undated photo, pulling his little brother Dean in a wagon. The two boys were inseparable growing up and were always known around town as "Dougie and Deanie." DEMOE FAMILY

Jerry DeMoe in 1966, the year he graduated from Tioga High School. He went to live with his older brother's family when their mother, Wanda, was committed for symptoms they later learned were caused by Alzheimer's disease. SHARON DEMOE

Sharon Bratton, Tioga High School graduation photo, 1966. She would marry Jerry DeMoe, her high-school sweetheart, two years later. SHARON DEMOE

Jerry and Sharon DeMoe on their wedding day in 1968.

SHARON DEMOE

Moe's sister, Pat, settled in rural Wisconsin with her husband, Rob Miller, and their three daughters. Pictured here in 1969, left to right: Colleen, age 5; Robin, age 6; and Dawn, age 7.

REBECCA VORK

Gail and Moe's six children, roughly 1971. Left to right: Dean, Karla, Lori (holding Jamie), Doug, and Brian. DeMoe family

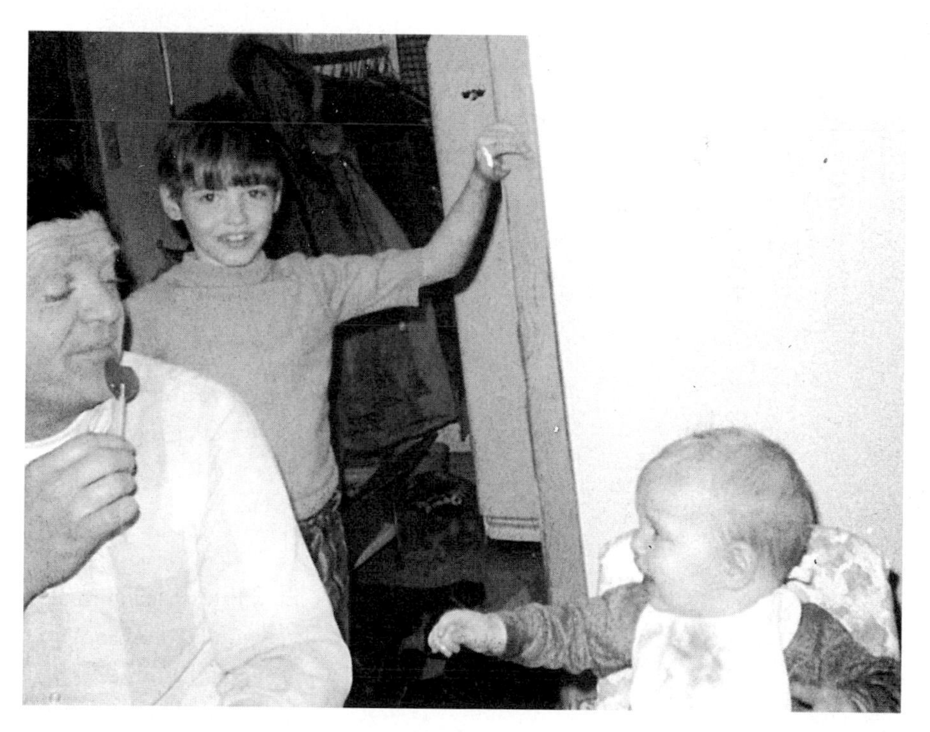

February 1972: Moe with his sons Dean (standing) and Jamie. A year later, prompted by her growing concerns about his disorientation and forgetfulness, Gail would bring Moe to a neurologist. DeMoe family

Gail visiting Moe at the VA hospital in St. Cloud, Minnesota, where he lived from 1978 to 1986. He asked for her and the children often and pleaded to come home.
DEMOE FAMILY

The DeMoe siblings, 1986. Back row, from left: Jamie, Doug, Dean, and Brian. Front row: Karla, Gail, and Lori. By now, most were starting families of their own. Though their father would die three years later from autosomal dominant Alzheimer's disease, none of them realized the seriousness of their own risk. DEMOE FAMILY affects everyday decisions, such as what to eat, whether to exercise, what kind of insurance to purchase. Do you train for that marathon? Keep trying to learn a new language? Quit the dead-end job that pays the bills? When is it appropriate to tell a romantic interest: The third date? Fifth? Tenth?

In families like the Noonans and the DeMoes, genetic testing for their mutations can predict with absolute certainty from any age, including infancy, whether a person has inherited the disease. In the general population, genetic tests offer less certain results but still carry a burden of knowledge that can be life-altering: a Faustian bargain.

Imaging from a PiB derivative such as Amyvid allows doctors to diagnose Alzheimer's in a living brain, but without meaningful treatment or means of prevention, giving someone such a diagnosis becomes an ethical land mine. In fact, outside of clinical trials, Medicare will not pay for such scans.

Even a clean bill of health carries drastic implications: How do you care for those around you who tested positive? People who get a negative test for the mutation aren't guaranteed a long life, either, as the Noonans would eventually find out.

It was that terror of losing the next generation that spurred the DeMoes into action. Despite Jerry's failure to respond to Cognex, the lack of details from their earlier attempts at research, and the knowledge of what kind of end was in store for those with a positive diagnosis, the DeMoes saw only one path toward hope: If their bodies could help science ferret out an answer, they might save their children. Viewed in that context, there was really only one choice to make.

It would be an emotionally expensive choice, even for those who didn't carry the gene. For Gail, the center of the family and life of the party, any positive diagnosis from the research was bound to carve years off her life in grief and worry. Only the belief that their sacrifices would save her kids' children—and other people's children—kept her from total collapse.

Karla and Dean decided they wanted to know their status. If they were going to assume responsibility for Brian and Doug, they reasoned, they should probably do so with a clean bill of health. The study personnel drew their blood, but before they were allowed to learn their results, genetic experts counseled them about the possible ramifications of that knowledge. Both promised to weigh the decision carefully.

Before they left, Trey Sunderland said he hoped to enroll Lori and Jamie in his study, in case they had the gene or could serve as controls for those who did. In Wyoming, Lori was worried about the possibility of inheriting the disease, but decided to apply for long-term care insurance before finding out her status. Jamie was more concerned with working and chasing women than he was with the possibility of Alzheimer's; his brothers just seemed so much older than him. Testing was something he eventually planned to do, but he didn't feel any sense of urgency.

Sunderland also put Karla in touch with Julie Noonan Lawson; for the first time, the DeMoes would be able to speak with one of the few people in the world who knew what they faced. For Karla, being able to share her fears, frustrations, and questions with a seasoned veteran was a gift for which she would be extremely grateful.

Eleven

WHEN THE FOG ROLLS IN

JULIE NOONAN LAWSON was five years old when she first became aware that her mother was starting to fade.

She was just a tiny thing, a kindergartener clad in a coat and hat to protect her against the harsh Boston weather, waiting by the school door for the mother who never remembered to show up. The eighth of ten children in an Irish Catholic family, Julie might have learned one day that such lapses happened occasionally in large clans; but the increasing absence of her mother was an omen that could not be explained away by absentmindedness.

There was a time when Julia Tatro Noonan whirled through her day, baking, singing, attending to her children, listening to their prayers. But her youngest daughter did not get to experience much of that mother. Rather, the mother she knew forgot to change her younger siblings' diapers, lost her husband's paycheck, struggled with the coffeemaker, overreacted to small slights. She was only in her late thirties; like the Italian woman Jean-François Foncin stumbled across in his Paris hospital, she had babies at home. It was the early 1960s, and Alzheimer's was still widely misunderstood. Her behavior baffled doctors.

Thinking Julia might be suffering from postpartum depression, they suggested time off. She took trips away from her family, but her symptoms only worsened over time. As was customary for some cases of the day, she was given shock treatments. Finally, she was taken to the psychiatric unit at a hospital, and in 1967, she was diagnosed with Alzheimer's disease. She was forty-three.

For the remaining eleven years of her life, Julia was institutionalized. Julie saw her mother strapped to a bed, medicated, wasting away. She died in 1978, her weight hovering somewhere around eighty pounds, curled in the fetal position, unable to recognize her children. Just like Wanda DeMoe.

The ten children Julia left behind were, in descending order, Dennis, Maureen, Patty, Kathi, Fran, Butch, John, Julie, Bob, and Eryc. After Julia disappeared from her family, her husband, John, was overwhelmed by the responsibility of caring for all those children. The youngest went to live with married siblings. For the next several years, the family fought to regain a state of equilibrium, reorienting itself after the loss of its mother. And then, just when they had started to normalize, the bottom dropped out again.

When Julia fell ill, her oldest daughter, Maureen—then in her twenties—and her husband, Dick Preskenis, adopted one of Mo's little brothers, four-year-old Bob—the ninth child in the family, born after Julie. He even took their last name, and he was raised as a sibling to their own children. In a cruel irony, Bob lost his second mother when Maureen also developed the same symptoms that had stolen their biological mother.

When Kate Preskenis was in eighth grade, her mother, Maureen, sat her down to watch a scratchy video of a made-for-TV movie called *Do You Remember Love*. It starred Joanne Woodward as an English professor whose decline from Alzheimer's disease affects her family and her colleagues.

Maureen cried through it. Afterward, she asked Kate if she had any questions, and explained that the family thought there was a possibility they had a genetic link to the disease because Kate's great-aunt Agnes, her grandmother Julia's identical twin sister, also had Alzheimer's.

Maureen and her siblings, who felt normal, theorized that the disease might skip a generation: She and her siblings thought they would not have to worry about it, but their children might. Fran's husband, a doctor, had attended a conference and returned with that suggestion.

But if the theory proved wrong, Maureen's viewpoint was chilling, albeit a common one for the children of Alzheimer's patients. If she developed the disease, she said bluntly, she did not want to live. She never wanted to end up like her own mother, curled in a permanent fetal position.

Her request was emphatic: Let me go.

Those words would haunt Kate for a long time.

About two or three years later, Kate began noticing small signs in her mother: She seemed aloof, her recollection of events wasn't as sharp. Within another couple of years, Maureen was repeating questions and having difficulty processing the answers. The situation worsened when her husband, Dick, died suddenly from a heart attack; he'd been under increasing amounts of stress, including that from watching his wife, who was his best friend, deteriorate.

Not long after Maureen's symptoms appeared, her sister Fran also began to decline. Fran was fifth in birth order and considered the family genius; she had a gift for technology and worked for IBM. Despite the eight years between them, she and Julie were particularly close.

Perhaps it was that bond that allowed Julie Noonan Lawson to become so attuned to changes in her sister's demeanor, things she would not have noticed in Maureen, who lived in upstate New York; perhaps, because Fran was so intelligent, her small blunders stood out in sharper relief against her everyday brilliance. Despite the family's theory that the disease would skip their generation, in 1992, Julie began to realize that something was wrong.

"I recognized her blank stare," Julie said. "And it was so intuitive that nobody around me agreed with me. Some of it has to do with our connection, and me being five years old and losing my mother."

In Fran, she sensed the same emptiness: Even when Fran was looking directly at Julie with her pale blue eyes, nothing seemed really to be reaching the soul behind them. Fran was diagnosed with Alzheimer's in August 1994, a week before her forty-fifth birthday.

Though they had had a few fleeting years of respite from the disease, the family realized, they hadn't escaped. "We all were kind of hellbound on 'OK, what are we going to do?'" Julie recalled. "'We can't just let this overcome us again.'" So when she heard about an Alzheimer's conference in Boston, designed for health care professionals, she tagged along with a friend who worked in the field and whose mother also had the disease.

One of the presenters was Trey Sunderland, who explained his protocol of spinal taps, blood draws, and brain imaging. Although everything he discussed was aimed at people who were fifty or older, Julie was immediately impressed. It was the most scientific approach to research that she'd seen yet. Then in her thirties, she strode up to Sunderland and bullied her way into his study.

"I know I don't meet your protocol," she explained, "I know I'm too young. I know my oldest siblings are just barely going to be in the window. But we've got it. My mother had it; my sister has it."

Julie and Karla DeMoe had a great deal in common, not the least of which was determination. Neither one would take no for an answer. Trey Sunderland gave her his card, and the Noonans began their own family's journey into the sometimes murky world of Alzheimer's research.

The Noonan siblings completed memory tests and gave blood samples at Massachusetts General Hospital. But the siblings were looking for more; more answers, more research, more opportunities to contribute. Like the DeMoes, they were determined to do everything they could, but before they would agree to have their body fluids extracted for the NIH study, they wanted a guarantee that the material would be shared with any other researcher who wanted it. Sunderland assured the Noonans that their samples would be available to others.

One of Sunderland's studies, which began in 1995, focused on a point in the Alzheimer's timeline that was on very few scientists' radars at the time: those years in middle age, well before symptoms typically develop, when amyloid and tau might be silently accumulating in the brain. In those years, nearly a decade before Bill Klunk and Chet Mathis developed their amyloid brain scans, Trey's protocol checked patients each year with the tools that were available to him at the time—an MRI, clinical and cognitive testing, and spinal fluid samples.

Marilyn Albert, a cognitive neuroscientist who now leads the Alzheimer's Disease Research Center at Johns Hopkins University, said Trey's study was way ahead of its time, particularly in its use of spinal fluid. Several studies now look at amyloid and tau that has spilled into the fluid cushioning the brain and spinal cord, and researchers theorize that the amount of these proteins—when measured along with brain scans and family history—can reasonably predict Alzheimer's in the general population. If a prevention treatment is to be found, refining various ways to forecast Alzheimer's will be key in deciding whether a person needs to start taking the medication, and when.

Beginning in 2000, the Noonan siblings went every year to donate their spinal fluid. They also completed sleep studies and all kinds of seemingly unrelated tests.

From year to year, their answers to simple questions—*Where were you born? What is today's date?*—would reflect the widening cognitive disconnect. Some were simple math problems, such as calculating the number of nickels in a dollar or subtracting backward from twenty in sets of threes. Others were everyday problem-solving questions: *If you were visiting a friend in an unfamiliar city and you needed to find out where they lived, how would you do it?*

In an effort to help their doctors better understand their disease, the family provided the NIH with the name of the institution that housed the autopsy results for their mother's twin sister.

Just as Maureen had said, both Julia Tatro Noonan and her twin, Agnes, had had Alzheimer's; although they were identical twins, their age of onset was dramatically different, with Agnes not showing symptoms until 1975, when she was fifty-one years old, more than a decade after her twin. She eventually died from lung cancer ten years later, when she was sixty-one. But an autopsy revealed that Alzheimer's had ravaged her brain.

Agnes never participated in research during her life, though in the 1970s, when both women were still alive, Alzheimer's research had not yet progressed enough to fully explore the rare scenario they presented. As identical twins, they shared the same genome, including their genetic mutation. But because their life spans differed significantly, researchers later theorized the difference in their disease progression may have been epigenetic, caused by outside factors that modify the genome and change the pattern of how genes are expressed without changing the actual DNA sequence. Diet and environmental factors, for example, can create epigenetic changes.

In 2014, teams from Brigham and Women's Hospital in Boston and the University of Exeter in England independently found hints of epigenetic changes in Alzheimer's patients, but emphasized that more research was needed to determine whether the changes accelerated the disease or happened as a result of it.

Maureen and her family relocated to Oregon. When she died in November 2001 and Julie broke the news to their sister Kathi, who was fighting a painful battle against breast cancer, Kathi simply said, "Good for her." At least her sister was finally free from the Alzheimer's curse. Kathi was hours away from death herself, but she considered herself better off than Maureen. She told Julie: "At least I still have my mind."

Maureen and Kathi died on the same day; Maureen on the west coast, Kathi on the east. The funerals were spaced far enough apart that Julie flew to each, remembering Kathi's words.

By June 2003, knowing that a test was available to find the APP gene—the rarer mutation, found on chromosome 21—Julie was ready to learn her own genetic status, which she was convinced might help her regain control over a situation that felt like it was spiraling out of her hands. When she had been hurt as a child, her father used to tell her: "Walk on it; it will get better." But the losses kept mounting—first her mother, then her sisters—until she staggered under the weight of the burden.

All she could think was: There had to be some reason for the pain, some end in sight. She struggled to articulate why knowing her mutation status would change that feeling, but her genetic counselor was able to explain:

"You are in a suspended state of life," he said. "You can't hold on to your past and everything you know from your past, because you don't know your future. Your future is suspended. And my guess is you will feel differently after you know."

Julie believes now that some of what he told her was standard operating procedure in genetic counseling; "but when he said it, it was like someone put an oxygen mask on me."

On June 19, 2003, her results came back: She was negative.

Of Julia's ten children, three have died from Alzheimer's: Maureen, Fran, and Butch. Of the five remaining children, four tested negative for the gene, including Julie. One declined to be tested. Two more children died from other causes: Dennis, the oldest, and Kathi.

A year after Julie was cleared, in August 2004, four of the DeMoe siblings met with Trey Sunderland at the NIH, and Karla subsequently contacted Julie.

"I felt like I knew her; I knew who she was," Karla said. They were kindred spirits, sisters struggling to make sense of a senseless disease that was choking their family trees.

For Julie, there was the same sense of validation, of comfort: *Our family is not alone. We are not freaks.* Although the two women have never met, their phone conversations have provided a level of support that nobody else has been able to offer. Like Karla, Julie feels a responsibility to pursue answers; as a survivor, she wants her siblings' deaths to mean something. Each is the narrator of her family's tale, keeping track of mounting losses, telling the story again and again in hopes of finding a way out. It is a lonely burden they share.

When Fran was still alive and well enough to speak, she testified before the US Senate's Special Committee on Aging in June 1995. Two decades after Robert Katzman's editorial call to arms, Alzheimer's had gained growing recognition as a major public health crisis. The committee had convened to discuss how increased research funding might pay off in saving long-term health care costs related to Alzheimer's and other brain diseases.

Just a week earlier, Jonas Salk had died. In his opening remarks, the committee chairman, Senator William S. Cohen, compared the effort to find effective Alzheimer's treatments to the fight to prevent polio.

When it was Fran's turn to speak, she told the lawmakers how she had once been the family's memory bank, the one who always knew schedules down to the last detail. Then she started having what she called "brain blips," when in the middle of a thought or a sentence, she simply went blank.

"If you have ever lived near the ocean, it's like when the fog rolls in," she said. "You can feel the emptiness inside."

She knew, even before she was diagnosed, what was coming. It was the worst feeling she'd ever had in her life. She told the senators of her shame when she forgot appointments and details; she didn't want the people in her life to feel unimportant. She hung dry erase boards throughout her house to help her remember, and she wrote with markers on windows and mirrors. She installed a car phone so her children, then fifteen and thirteen, could reach her if she forgot to pick them up.

"The only hope that I have is that my children, my family, and millions of others will never have to face this themselves," she said. "This depends upon continued research funding by this government so that a cure or prevention of this disease can be found. The government is attempting to cut back in all areas. If Capitol Hill cuts back in this area, they will pay much more later for what they could be taking care of now."

She had told her children she did not believe they would have to suffer through Alzheimer's twice, first as a caregiver and then as a patient, as she herself had. As long as the family had participated in research, they'd been given the same forecast: A treatment was coming, but it was probably five to ten years away. The irony was that, as years passed, that prediction never changed.

After Fran spoke, a research doctor approached her as her son and daughter stood by her side.

"I just want you to know, you do not need to worry about your kids," the doctor reassured her. "We will have a cure for this."

Eighteen years later, with Fran gone, and the prediction still "five to ten years," Julie reflected on that moment: "I'm glad she went down thinking that way."

But Fran's children, now grown, still live in a world with no cure.

Twelve

MALDICIÓN

"Incredible things are happening in the world. . . .
Right there across the river there are all kinds of magical
instruments while we keep on living like donkeys."
—Gabriel García Márquez, One Hundred Years of Solitude

THE CONNECTION BETWEEN Karla DeMoe and Julie Noonan through Trey Sunderland's NIH study offered comfort to two families who had previously been isolated by what seemed like a personal plague. What they did not know was that three thousand miles south of Tioga's flat plains, in the rural mountains of Colombia, lived a cluster of people who shared their fate. If science could find a way to merge them in research, they would have an ideal group for testing possible drug therapies.

Dr. Francisco Lopera has taken a long, complicated odyssey through the world of Alzheimer's, persevering against much greater odds than most of his American-based colleagues in his quest to find a treatment for his people. In many ways, Alzheimer's is like the guerrillas who populate his native Colombia: Sometimes it lies in wait, and sometimes it hides in plain sight, but it is always deadly, always calculating, and nobody is sacred. When as a young doctor

Lopera met his very first Alzheimer's patient, he had no way of knowing that his efforts would hold the key to uniting global efforts to find a treatment.

In pursuit of his enemy, Lopera has trudged along winding one-lane dirt paths in the rural outposts of Colombia, as treacherous a terrain as it is beautiful, where kidnapping is so commonplace he has matter-of-factly asked colleagues to bring toothbrushes when they accompany him—just in case. As a young doctor, Lopera had himself been kidnapped and taken deep into the forest to treat a wounded guerilla soldier, then returned to his hospital. Though he had grown indifferent to the danger, the volatility of his country would significantly hamper his efforts to collaborate with other Alzheimer's researchers.

A childhood fascination with flying saucers originally drove Lopera toward astronomy. But after reading a newspaper article that said flying saucers existed only in the mind, he thought, "If that is correct, it is better to study the mind." At university, he started off in psychology, then switched to neurology because it was a stronger program. If he studied the brain, he reasoned, the mind would soon follow.

After serving an initial residency in the rural outposts of Choco, Lopera moved to Medellín, where he was completing a residency as a clinical neurologist, with a special interest in the relationship between language and the brain.

One day in 1984, relatives brought to the clinic a forty-seven-year-old man suffering from almost total memory loss, rendering him unable to work. He had been experiencing his symptoms for four years, his family reported; now he had deteriorated to the point where he simply stared into space or burst into maniacal laughter.

"I was shocked by that, because he was very young, and because his father and his grandfather had the same history," Lopera said.

While Lopera was gaining new insight into the possible heritability of the disease, back in the United States, George Glenner was floating the idea that a specific defective gene was triggering Alzheimer's, though it would be another three years before Dmitry Goldgaber would discover APP. Karla

DeMoe was a new mother, and her father, Moe, was just reaching the point where he had forgotten his own children's names.

Lopera followed the afflicted man to his town, Belmira, and talked to other relatives, some of whom blamed the illness on an ancestral curse, known as *la maldición*. With their help, he reconstructed the history of other cases of dementia among their relatives, and he created the first pedigree for that group. *La bobera*, they called it: the foolishness. Some said a village priest had caused the epidemic by casting a spell on parishioners who stole from the collection box. Some said people lost their minds if they touched the bark of a particular tree.

In the United States, members of families that were similarly affected by Alzheimer's also invented rationalizations for the disease. Some thought it affected women more than it did men (this is supported by statstics: almost two-thirds of Americans with Alzheimer's are women, partially owing to the fact that they generally live longer). Some thought you were more likely to get it if you physically resembled a relative who had it, and Brian DeMoe's lookalike son, Yancey, was one of them. Some tried to decipher patterns of inheritance, willing the disease to skip a generation, just as the Noonans once did.

Lopera summarized his findings in a paper and published it three years later in a Colombian medical journal. From that point forward, his career would be devoted to solving the mysteries of autosomal dominant Alzheimer's disease.

The obstacles Lopera faced were formidable. Lucía Madrigal, a veteran member of his research team at the University of Antioquia, was cast from the same mold as hard-boiled battlefield nurses everywhere. At times during her career, she has had to make her rounds by horse or on foot; as roads became safer, she was able to reach some families by car, as June White had been doing since the 1980s.

On one trip, while collecting blood samples in Angostura, Lucía was kidnapped by a drug cartel and held for eight days. She ordered them to take care of the samples, regardless of what they did to her. Surprisingly, the drug lords obliged, keeping the vials cold in a river until she was released. A few months later, she returned; one of her kidnappers, a guerrilla boss, asked her to help with his mother, who was stricken with Alzheimer's. She agreed.

"We had a period very, very dangerous for us," admitted Lopera. But he was quick to add: "The violence is better now."

By 1992, eight years after his first Alzheimer's patient walked in, Lopera had identified three extended branches of the family in the state of Antioquia, a region in the Andes Mountains, who had the disease. As word of his interest spread, more people came through referrals. They had all kinds of problems: lost memory, difficulties with perception and cognition; some were given to outbursts and fits. It seemed *la maldición* manifested itself in all sorts of ways, and in its own time; some were afflicted in their late forties, one was symptomatic at thirty-two. But in the end, the disease caught up to each of them the same way.

Kidnappers weren't the only unusual source of Lopera's research leads. His driver, Antonio, would sometimes tip him off to new patients he'd discovered on his own, asking them the same questions he'd overheard the medical team asking. In an environment unencumbered by health care privacy laws and attorney-drafted release forms, everyone sat around the family's table to discuss the finer points of the disease: the patients, the doctors, the psychologists, the driver.

Before he had a computer, Lopera kept track of Alzheimer's victims on index cards; when he couldn't reach them by car, he—like Lucía—traveled by horseback. Despite his efforts, he worked largely in obscurity for decades, hampered greatly by the violence and isolation of the corner of the world in which he lived.

What Lopera had stumbled upon was the largest known family afflicted with the genetic mutation form of Alzheimer's—in their case, PS1. For centuries they had been living largely in towns scattered throughout Antioquia, isolated from the rest of the world by geography and the warring factions that kept outsiders at bay. During the years he studied them, he mapped out the family tree to extend to five thousand members, all descending from Javier San Pedro Gómez and María Luisa Chavarriaga Mejía, a couple whose records dated back to 1745, and who settled near

Angostura. Lopera believed about 30 percent of the family—or fifteen hundred people—carried the mutated gene. Outsiders knew the family collectively as the *paisa*, which is also a general nickname for the people of northwestern Colombia. Their genetic makeup—a mix of European and Middle Eastern conquerors and immigrants who blended with native people and then were isolated by the mountainous geography—was its own unique cocktail, and in the family of Javier and María Luisa it hid DNA-encoded secrets that could potentially affect the Alzheimer's population worldwide, for the family offered the most perfect group of research subjects that science had yet found: from the same general environment, with the same mutation, in large enough numbers to make a study scientifically feasible.

Now all Lopera needed was to connect what he'd learned with the researchers who were working with similar mutations elsewhere in the world.

Lopera left Colombia for two years to attend Catholic University of Leuven in Belgium for training in behavioral neurology, taking with him the family pedigree that he'd so painstakingly constructed. He was certain that in Europe, he would find other scientists who shared his excitement over the discovery, but he was disappointed. Nobody seemed to recognize the importance the family represented. He returned to Colombia with his index cards and waited for a better answer.

Breakthrough came in the form of Ken Kosik, a young Harvard neuroscientist with a literary mindset. It was life that really interested Kosik, not simply brain theory. Like Lopera, he cared about the mind.

Before he attended medical school, Kosik earned a master's degree in English literature. He saw his interest in both brain science and literature as complementary obsessions, and often told his students that the humanities offer good training for science.

"I think the things that make people curious about the brain are all in literature," he said. "If you only train as a scientist, then you sort of get the impression that everything in the world is either black or white, right or wrong, or there's a very clear answer."

Take, for instance, the neurofibrillary tangles that are found in the brain cells of people with Alzheimer's disease. The chief ingredient in tangles is the tau protein, which is also found in the normal brain. Tau is soluble, meaning it dissolves in most liquids. But when it becomes a tangle, the opposite happens: Suddenly, it becomes so insoluble that absolutely nothing can break it down. For many scientists, this dichotomy presented a baffling paradox when it was first discovered. But Kosik was comfortable with the idea of dual realities and the puzzles they present; he wanted to analyze them.

Kosik was working in Cambridge, Massachusetts, when he struck up a friendship with a neurosurgeon from Colombia. Together they got a small grant of a few thousand dollars to promote neuroscience in that troubled country. In October 1992, they traveled to Bogotá, the capital, where Kosik gave a lecture on Alzheimer's disease. Lopera was in the audience.

After Kosik finished speaking, Lopera introduced himself and told him about the families he'd been studying. Kosik had read Lopera's paper in the Colombian medical journal, so he understood some of what Lopera was arguing: that this family was valuable, offering untapped potential for research.

With the dawning realization that Alzheimer's was actually a much bigger problem than anyone had previously realized, the public was beginning to scramble for answers, and science did not have many to provide. Kosik had learned to guard against people who reached out to him—even fellow scientists.

"I said, 'OK, tell me more,'" Kosik recalled. "I wasn't really sure; because when you're in the Alzheimer field, many people tell you things: 'My mother has Alzheimer's. My father.' You have to sort of discriminate."

Lopera produced his index cards documenting the family tree, the ones everyone in Europe had ignored. But to Kosik, they were intriguing; it seemed clear that Lopera had uncovered something groundbreaking. Goldgaber's APP gene, while exciting news, only accounted for a small percentage of the known families with Alzheimer's. With PS1 yet to be discovered, researchers—including Kosik—knew something else was out there, and a large, as yet untapped resource like Lopera's family represented a huge opportunity to try finding another gene.

Recounting that moment, many years later, Kosik would write that what occurred next was the sort of thing that only occurs "at a stage of life when very little prevents us from following our instincts and the allure of the unknown." The next morning, he boarded a plane to Medellín.

They started in the mining town of Yarumal. At first, the Lopera team was exceedingly cautious about taking an American doctor into the notoriously violent region. Two bodyguards were assigned to Kosik, who shrugged at the risk.

Kosik quickly saw how beloved Lopera was to the people he was studying. When he arrived, they threw open their doors, inviting all their relatives to see him, offering food and friendship. Kosik was struck by how warm they were, with the doctors and with one another; in the midst of death, they continued to celebrate life.

While Lopera conducted a basic neurological exam on a man in his late forties, Kosik watched the doctor hop on one foot, demonstrating what he wanted his patient to do. The man obliged. But once he began hopping on one foot throughout his small house, he didn't understand that it was time to stop; as he hopped, his young son and niece giggled, delighting in their favorite playmate, and unafraid of his disease. Lopera sat to speak with the man's wife, whose sixty-three-year-old mother—also stricken with Alzheimer's—lay in an adjoining room, unresponsive to Lopera's gentle cues. She was frozen in the crooked posture of an end-stage Alzheimer's patient, but her family continued to bustle around her, touch her, speak to her.

Kosik was moved by this world without nursing homes or hospice, where children still held the hand of a grandmother who did not know them, where family members embraced their dying instead of isolating them until the inevitable, early end. He came away thinking there was much to learn from the *paisa* and their attitude toward the afflicted, lessons that transcended even the scientific clues their brains and blood and spinal fluid might yield. He wrote about what he saw, but his true focus was on finding the source of the disease.

Back in the United States, the quest to find genetic mutations was well under way. But first, Kosik and Lopera had to confirm that the *paisa* disease was definitively Alzheimer's and not some other disorder. It was the mid-1990s, and Klunk and Mathis were still working on developing PiB. The only way to absolutely confirm such a diagnosis was by looking at slivers of brain tissue under a microscope. The problem was that Colombian funeral rites were rife with superstition, and so far, none of the families had agreed to allow Lopera's team to take a loved one's brain for analysis.

Among Lopera's patients was a middle-aged mother of fourteen children who had been battling her disease for eleven years. When Lopera learned that she had died, he and his colleague Juan Carlos Arango, a neuropathologist from the University of Antioquia School of Medicine, drove five hours to the pueblo in Angostura where the corpse of Lopera's now-deceased patient was laid out in her sitting room.

The woman's body was surrounded by family, friends, and the *lloronas*—professional mourners. Lopera and Arango prayed and sat with the group as they drank, talked, and cried through the night. And as they talked, thirteen of the dead woman's children eventually agreed to Lopera's request; they would turn her brain over to science. The fourteenth resisted, and as the night progressed, he became more belligerent.

He demanded payment for his mother's remains. A onetime policeman who was rumored to work as a hit man for a drug cartel, the son was sure the doctors were going to sell his mother's remains to the gringos. As they waited, across the room the woman's brain continued its rapid decay until it was on the verge of being scientifically worthless.

Finally, inexplicably, the son relented. Wasting no time, Lopera and Arango raced with the body to the local infirmary. Working as quickly as they could, they deftly removed the brain and dropped it into formaldehyde, packed it in Arango's carry-on bag, and got him on a plane to Boston. There, in Kosik's laboratory, Arango proved what the team had believed all along: They were dealing with a particularly insidious form of Alzheimer's. It would be a turning point in the *paisa*'s story, confirming their potential value to the research field.

As the research continued to evolve during the next few years, Lopera's and Kosik's relationships to their subjects grew deeper. The people willingly shared the most personal details about their lives; without that frankness, Lopera's work would have been impossible, because the anecdotal details supplemented his scientific findings. Defying centuries of superstition and religious teaching, the *paisa* began to donate the brains of their loved ones to the research more frequently; the ravaged organs sat in round white plastic tubs, labeled in black Magic Marker, one of the largest banks of its kind and a priceless tool for science. The doctors who sliced and microscopically analyzed these brains were the same people who sometimes broke bread with the victims when they were alive and witnessed their decline as the disease took its inexorable hold. The kinship made the scientists more fully invested in their work. It became more than a fascinating scientific problem; it was personal.

"You are *paisa*," Lopera told Kosik one day, many years into their collaboration. Kosik considered this one of the greatest compliments he had ever been paid.

To aid their efforts in identifying a new gene, they teamed up with Alison Goate, a geneticist from Washington University in St. Louis. They were hard at work when Peter St. George-Hyslop, the neurologist and molecular geneticist who had teamed up with Jean-François Foncin on Family N, identified the PS1 mutation—which affected a wide swath of genetic Alzheimer's families, from the DeMoes to the *paisa*.

Lopera's team remained frustrated in their attempts to get the outside world to notice their hard work. The remote location that had isolated the *paisa* mutation for three centuries, allowing it to proliferate among cousins who married cousins and bore a dozen children, also made it difficult to get the attention of the medical community.

Even Kosik, who was embedded in Alzheimer's research in the United States—in Boston, no less, a virtual mecca of scientific research—could not convince anyone to study the Colombian family more deeply. Unlike Family N, which was rooted in Europe, the *paisa* occupied a part of the world that many people considered too dangerous for the kind of extensive travel necessary for the job.

"I have been in the Alzheimer field a long time. I talked to my colleagues about [the] Colombia project. They don't listen; they don't believe," Kosik said. "People didn't want to hear about research in Colombia. It was too far away; it was too dangerous. The Alzheimer field is very closed, and they talk only to themselves. Because we work in obscurity for twenty years, for Dr. Lopera even more. And no one really cares."

In the meantime, as the world ignored them, the families waited.

In the United States, the DeMoes and the Noonans had the difficult but available option of choosing to know their fate. Some did, and some didn't. Genetic counselors guided them through those decisions, gave them variables to think about and advice that helped each person make this very individual choice. But in Colombia, there were no genetic counselors; and without them, or a means of preventing the disease, doctors deemed the knowledge too dangerous to share. Lopera didn't want to play God.

The differences between American and Colombian cultures also influenced the amount of autonomy patients had over their medical information. Americans experience a subtle stigma associated with Alzheimer's: for example, life insurance can be harder to obtain. Among the Colombian families, the vein of stigma and despair associated with the disease is more overt, with people openly vowing suicide if their memory failures turn out to be Alzheimer's—a sentiment quietly shared by many Americans. For that reason, doctors do not disclose the patients' genetic test results. In the United States, where people are accustomed to controlling access to their own medical information, they have more leeway to demand answers.

Such is the nightmare of Alzheimer's researchers, who must navigate an ethical tightrope between honoring a patient's right to learn his or her own status and safeguarding the patient from potentially lethal knowledge. Kosik recalled how he and Lopera once met with a family in a Medellín barrio and asked them what they would do differently if they knew their test results. The only answer came from a twenty-three-year-old man named Gonzalez, who said if his test were positive, he would shoot himself.

"We stand poised to be expelled from an Eden of genetic ignorance into a society where every talent and weakness, every wrinkle and freckle may be predicted from our genomes," Kosik wrote of the experience. "When the genes tell a cruel story, we must be prepared for the power and danger of that information."

Lopera and Kosik would one day share that burden with other doctors, but it would be several years coming.

One of the most surprising discoveries Lopera and Kosik made in working with the population was an eleven-year-old girl who inherited two copies of the mutation—one from each parent, known in genetics as a homozygote. Because the mutation in itself is so rare, finding a person with two copies was such a statistically remote possibility that nobody really expected it to happen.

The eleven-year-old was the first, but by November 2014, five more homozygotes had been located in the *paisa*, underscoring how unique the family was to Alzheimer's research: It was the only place in the world where even one of these people had been found.

In addition to the girl, the homozygotes included four women and one man, ranging in age from twenty-seven to forty-six. Kosik and his team reported their case histories in aggregate to preserve their subjects' anonymity.

The doctors had never run the mathematical calculations to predict how many more they'd find in the complex family tree; pregnancies with two copies of the mutation were not expected to survive.

"We know one thing for sure now: that having two copies of the mutation is consistent with life," Kosik said, which was, in itself, an important discovery.

The *paisa* homozygotes were defying steep odds. Though the little girl had mild mental retardation, she was relatively normal—in Kosik's words, "a very charming little girl." The doctors weren't sure what to expect from her future, or how it would compare to that of her relatives with a single copy of the mutation.

"She's going to get the disease, and maybe she'll get it earlier," said Kosik.

Back in Colombia, the girl and her parents remained unaware of her genetic status—or of the fact that she had redefined medical knowledge.

What Lopera and Kosik hoped was to find a way to prevent the disease before it began to destroy neurons. A prevention—though still a huge hurdle—was still easier than a cure for people who already had symptoms, since science had yet to find a way to resurrect dead neurons. Even a treatment that could delay the onset of symptoms for five years would be considered a major breakthrough.

It took the rest of the Alzheimer's field a long time to reach that conclusion, Kosik said. For years, trials focused on treating people who were already suffering from memory loss and other hallmark symptoms. Drugs famously failed to do anything to achieve those goals. They didn't work in Jerry or Moe, for example.

But when scientists gradually began to think more in terms of preventing the disease before it ever began—thus removing the obstacle of replacing dead neurons—the question became this: *What happens if we give these same drugs to people before they get the disease, but who are guaranteed to get it if we do nothing?* And the obvious follow-up question became: *Where do we find enough people who match that description to run effective tests?*

The answer: In Colombia. And North Dakota. And Massachusetts, and in other small pockets of the world where this tiny segment of the Alzheimer's population had been identified.

Thirteen

BURDEN OF PROOF

IN NORTH DAKOTA, threads were beginning to pull loose from the carefully mended fabric of Gail's family. When the lab sent Brian's and Doug's genetic testing results, the accompanying letter also recommended genetic counseling for the rest of their relatives. Karla swung into action, contacting John Martsolf, head of the Division of Medical Genetics at the University of North Dakota's School of Medicine and Health Sciences. For the sake of efficiency, they set up a single meeting with him for the extended family in August 2004, inviting all the siblings, siblings' children, cousins, aunts, and uncles they could find. Karla was thorough, arranging to videotape the meeting for anybody who couldn't attend. Jerry's widow, Sharon, and their two daughters in Oklahoma were among those who missed it.

Among those who did make the trip were Moe's sister, Pat, now sixty years old, and her three grown daughters, Dawn, Robin, and Colleen—Gail's nieces, and cousins to the DeMoe kids—all of whom still lived in northern Wisconsin and now had families of their own. For three years, Pat had been exhibiting signs of Alzheimer's disease—much later than her brothers Moe and Jerry. At the meeting, Pat saw herself in a monitor that was connected to the video camera Karla had brought. She squealed in childlike delight: "I'm on TV!"

Dawn, the oldest of Pat's daughters, enjoyed a close connection with her cousin Lori McIntyre. Both were creative, lively women who had the knack for making friends wherever they went.

At the meeting with Martsolf, Dawn asked insightful questions. Since her midthirties, she had developed a habit of repeating herself. Her moods were starting to shift dramatically, too, straining her relationship with her daughters. Doctors had prescribed antidepressants, but her symptoms weren't improving. Now she was forty-two, and her work as an architectural draftsman was beginning to suffer. She was looking for answers.

Steve McIntyre made the trip to gather information, but Lori did not. After learning of her brothers' diagnoses, a nugget of fear had settled in the back of Lori's consciousness. Though she had taken the practical step of applying for insurance to cover her long-term care if necessary, neither Steve nor her daughters noticed any changes. Still, she continued to quietly worry, and her fears were starting to erode her usual upbeat confidence. Of course, her self-perceived "symptoms" could simply be aftershocks from the stress of learning about Brian and Doug.

But in her heart, she didn't think so.

Eight months before the family meeting, she had called her three best girlfriends, all still living in Nebraska, and suggested they meet in Cheyenne for a slumber party on New Year's Eve. The occasion was vintage Lori, replete with party hats, noisemakers, and belly laughs. Her best friend, Robin Tjosvold, took photos of them cutting up with the other patrons at the hotel bar: for one night, they were neither mothers nor wives, just four women living in the moment, acting like the teenagers they'd left back at home.

The next morning, when everyone was awake, Lori went into the bathroom, pulled out the hotel-supplied box of Kleenex, and dropped it in the middle of the bed. She gathered her friends in a circle and announced the real reason she had called them together: "I have to talk to you about something."

And then she confessed what she feared: She had Alzheimer's disease. The memory of her father's descent into madness weighed heavily on her mind, and she was worried about the three daughters back home who would soon be adults—fearless, she hoped, but still fragile, still in need of their mother. She asked her friends to make a vow that they would do for her what she, had the circumstances been reversed, would have done for any of them.

"If I do have it, and it's going to run the course it had with my father, then I need you guys to promise me that you will be mothers for my girls and grandmothers for the grandchildren that I don't have yet," she said. "You have to promise me those things, because they will need you."

Robin was skeptical. A nurse with experience working in hospice care, she hadn't seen any changes in her friend; she thought perhaps Lori was simply emotionally overwhelmed. But the women, sitting in a hotel room on New Year's Day 2004, vowed that they would do as she had asked.

Just as they had throughout their relationship, Steve and Lori moved again—this time to Alpine, Texas, for the start of their youngest daughter's sophomore year of high school—to follow Steve's Union Pacific railroad job. Their daughters Jessica and Robin, now old enough to live on their own, stayed behind and began their adult lives as workingwomen. They decided to move to Denver, where they talked about one day opening a beauty salon together.

At first, Texas was like every other state the McIntyres had lived in: an adventure. Lori served as a companion for a fussy elderly woman who needed someone to drive her around on errands. The woman drove her own daughters crazy, but she and Lori got along. Like Gail, Lori had a warm, empathetic personality that drew people to her.

"We clicked," Lori said, though she had nothing bad to say about the woman's daughters. "It's different when it's not your mom."

Later, she went to work at the high school as an aide for the gym teacher, setting up equipment—the same routine, every day. Steve remembered this detail distinctly whenever anyone asked him when he first noticed that Lori was laboring with her memory.

He didn't notice anything different, but the gym teacher did. Lori could not remember the same instructions from one day to the next. It was a strange disconnect, because at home, she was as self-sufficient as always. Steve wasn't sure what it meant.

The year after the meeting with Martsolf, Dawn went to the Mayo Clinic and learned that she was positive for the mutation. She had been visiting a friend in downstate Wisconsin, and she got lost on the way home. What should have been a four-hour trip took ten. Worse, a state trooper had found her driving the wrong way on Interstate 84, against ongoing traffic.

The Mayo Clinic reported that the extent of her brain damage was much worse than she had anticipated; her forty-three-year-old brain was equivalent in size to an eighty-year-old woman's.

Dawn's college-age daughter, Leah, was furious at how long the health care system had taken to diagnose her mother.

"She had been going for years to her behavioral health doctor. She'd been going on for years saying, 'Something's wrong. I have this in my family.' And no one would put it out there as 'You're sick,'" Leah said. "It was for years. She was trying to tell somebody."

Dawn's mother, Pat, also had the disease, though she was almost twenty years older when her symptoms began to show. But Dawn was so young that she was misdiagnosed as depressed. It was a common mistake in early-onset patients, because the diseases shared many of the same symptoms: memory loss, impaired concentration, interruptions in sleep patterns.

Dawn wasn't the only person affected by the misdiagnosis. Her marriage crumbled under the strain of her mood swings. Her daughters were robbed of their mother for much of their childhood, often left to fend for themselves when Dawn was incapacitated. Her extended family offered little support to help any of them weather each crisis.

Still, Dawn was hopeful. After she finally learned it was Alzheimer's, the doctor reported that science seemed close to finding a cure. The idea bolstered her confidence. When she started taking Aricept (the trade name for donepezil, a widely prescribed drug meant to alleviate Alzheimer's symptoms), she announced that she felt smarter. That matched the findings of a study that would be published in 2012 suggesting that patients taking Aricept would perform cognitively better for about a year longer than patients who took a placebo.

Her younger daughter, Alayna, was skeptical that her mother could be cured: "How do you grow a brain back?"

After Brian and Doug were diagnosed, Dean and Karla agreed that as part of their pact to oversee their brothers' care, they would share power of attorney so they could jointly make difficult decisions. Remembering the toll the disease had taken on Gail's emotional health when their father deteriorated, they wanted to alleviate her burden. Since they were both still in North Dakota, where they could physically assist with Brian and Doug, their offer seemed to make sense. Gail was grateful that her children were willing to shoulder the burden.

At Karla's urging, Dean agreed to fly to Bethesda to participate in Trey Sunderland's study at the National Institute of Mental Health, which compared biological samples such as blood and spinal fluid to people's symptoms and needed controls in addition to active subjects.

The biggest takeaway Karla brought back from her first visit with a genetic counselor was what a positive diagnosis of the mutation could mean for her children, Amber and Cole. They were her greatest joy, such bright, happy kids, so focused on their family. They were both in college now, and starting to plan futures; Cole was training to become a golf pro, and Amber was training to become a physician assistant. Karla simply could not stand not knowing if she'd somehow saddled them with faulty genes. Finally, she called Sunderland and told him she was ready to learn her status.

In October 2004, two months after her initial blood draw, Karla returned to Bethesda, this time just with Matt. They tried to distract themselves with sightseeing but were too restless and worried. They had decided not to tell either of their kids where they were going or why. If it was bad news, they'd wait until the college year was over, then break it to them over the summer.

The next day, Trey Sunderland sat down with Matt and Karla, who were sitting together like the high-school sweethearts they had been. They listened quietly as Sunderland began unspooling his usual speech about what they should expect if they learned Karla had the mutation. Finally, he interrupted himself and asked if Karla would just like to know.

"Yes," she said.

She was free.

For the rest of her life, she would never forget that moment, never forget the relief and joy that washed through her, knowing that her children, and their children, would forever be safe from this unbearable inheritance. Cole and Amber had just been given the greatest gift of their lives.

Sunderland and the staff were elated to give her the news. Karla and Matt were euphoric. Finally, after so much fear and desperation, they had something to celebrate.

They left messages for everyone they could think of. When Cole answered the phone in his dorm in Michigan, he was surprised to hear his mother sobbing about the good news—he hadn't even known she was away, let alone where she was.

The next day, Karla and Matt flew back to Fargo, arriving early enough that they both went back to their jobs to finish out the workday. Karla's coworkers surprised her with balloons and a cake frosted with the inscription: *Now your only excuse is you're blonde.*

Inspired by his sister's celebration, Dean told Sunderland he also wanted to know his genetic status and flew to Bethesda in October 2005 to learn the results. Since he and Karla were already making plans to assume legal control of their siblings' care, he figured he might as well make it official.

Deb and Dean sat impatiently in Sunderland's office, Dean in a distracted fog, waiting for the doctor to deliver their moment of truth.

"You just want to know, don't you?" he asked, as he had with Karla. And then the bomb dropped: Dean had tested positive.

Deb remembers crumbling from the shock. How could he have it? There wasn't a thing wrong with him; he was clever; he was different. He worked with his brain as well as his hands. He didn't process information the same way as his brothers.

Her mind screamed at the improbability of what she'd just heard.

At first, Dean did not speak. His thoughts remained purely his own. He was a son of Galen DeMoe, a veteran of the hell that was this disease. He refused to allow the news to touch him.

He would not pity himself. There were people counting on him. His children; his wife.

He was a man who provided. He was a man, period.

Finally, he turned to his wife and said something she would repeat many times: "Deb, it's not what you're dealt in life. It's how you deal with it."

But wasn't this situation different? she wondered.

"To me, you can apply that to anything," Dean said.

Years later, she would still cry as she recalled how stoically her husband reacted to the worst news of his life. His first instinct was to reassure her. He took her to the building's fourth floor, where pediatric cancer patients were fighting desperately for survival.

"You want to feel sorry for me? Look at those kids who aren't even three or four and know they're terminally ill, and don't even have a chance," he insisted.

Later, she would write down what he'd said, and draw strength from it.

Looking back, she said, "I knew, right then, he was going to be just fine."

They returned to their motel room, knowing that the whole family was waiting for the celebratory phone call. They'd all been so confident that Dean had escaped the mutation. The disease had fooled them all. How could the two of them even begin to explain that mistake?

They told Karla first. It was morning; it felt like the last day of the world. Someone drove her home from work. *Her mother.* She didn't want Gail to be alone when she heard the news. She called one of Gail's friends to go to the house to be with her when the call came from Deb and Dean.

Like Dean, Doug had said nothing immediately after his own diagnosis. But when Doug learned about his brother, his closest sibling, he wept inconsolably.

Then Karla went to her bed, where she stayed for days. Lori was so far away; Jamie was so young. There was no one to help her. Her beloved younger brother was going to leave her. She was alone.

After a few difficult phone calls, Deb and Dean stopped calling people. Sunderland had been right to make them come to Bethesda to learn the diagnosis; such news should not be delivered over the phone. They decided to head home and tell the kids in person—their children, so loved, the center of their lives. Now they, too, were in danger.

Sixteen-year-old Tyler was playing in a basketball tournament that week, while nine-year-old McKenna stayed with her aunt's family in Thompson. When the kids were finally home, Deb and Dean sat them down. The room went silent and tears welled in everyone's eyes as Dean revealed the news.

Lindsey Sillerud, Dean's older daughter, was attending college in Bismarck. Though she and Dean had stayed close, she had not yet heard from him when she opened her email and noticed a message from her cousin, Karla's daughter.

I'm so sorry to hear about your dad's test results, it read. *I really didn't think he had it.*

Lindsey sat in shock, looking at the electronic shimmer of the letters on her screen. Karla had broken the bad news to her own kids before her father had ever had a chance to tell her.

"One thing with the DeMoe family: Once you tell somebody one thing, it goes," she said later. "It spreads like wildfire."

She called Dean, who apologized.

"That's not how we wanted you to find out," he said. He had wanted to tell her himself, in person, and not by phone; it hadn't occurred to him that she might find out some other way before he had the chance to do that. As with Tyler and McKenna, he tried to reassure her that everything would be fine.

She was breathless. It was the first time she had ever imagined something bad might happen to him.

Two nights later, while Dean was fast asleep, Deb heard a noise that woke her. Downstairs her nearly grown son was sobbing. She went to him.

"Baby," she said, and gathered him into her arms, his long legs draped over hers, and let him cry. They talked for hours, all through the night. That was the only time Tyler DeMoe really broke down, thinking about his father.

Fourteen

THE RISE AND FALL OF GOLDEN BOY

FOR BOTH THE DeMoe and Noonan families, Trey Sunderland had felt like a godsend. Not only was he empathetic to their disease, but his role at NIH gave him real power to help them; he was heavily involved in research that seemed to be truly meaningful. At last, they felt as though they had found someone they could trust, who offered them hope. He was beloved at the NIH as well—his office nickname was Golden Boy—yet Sunderland's days as the families' savior would prove to be numbered.

Depending on whom you ask, Sunderland either got what he deserved or was crucified for doing exactly what many NIH scientists, before and since, have always done: bridge a gap between public and private research to marshal all available resources in the fight to solve a public health crisis.

At first glance, few people seemed less likely to fall from grace. He won NIMH's Exemplary Psychiatrist Award in 2000, the NIH Director's Award in 1998, and a 1990 Public Health Service Commendation Medal. He headed NIMH's institutional review board, which governs the agency's ethics, for ten years.

But there was another side to the Golden Boy, one that yearned for recognition from the upper echelons of the Alzheimer's research community. In the end, his ambition would be his undoing, and bring down with him years of painstaking contributions from families like the Noonans and the DeMoes.

As a government agency, the National Institutes of Health aimed to objectively assess treatments without concerning itself about profits or losses. To NIMH director Thomas Insel, the NIH was "Camelot," and its internal research program—the largest biomedical research institution in the world—was the crown jewel. "It is a place for highly innovative, exciting science," he said.

As such, it was critical to him that its scientists were above reproach: "It is not good enough to just be clean," he said. "It has to be the place where no one will have any question about conflicts of interest. There has to be some place in the United States where the public knows that there is no taint, that there is no question, that there is no outside investment that is involved, that this is being done for the public good. This is the place."

The problem was that pharmaceutical companies did sometimes provide funding to NIH studies, and though theoretically there were boundaries to safeguard against any conflict of interest—such as the Institutional Review Board, to which scientists submitted potential conflicts—those lines tended to blur. "The NIH is supposed to be above all that," as one former employee pointed out. "In taking money from the companies, it really undermines what the NIH is supposed to be about."

At the same time, a lot of government scientists were consulting for private companies. Such arrangements, too, were permissible, provided they were fully disclosed to NIH leadership. There were caveats to those deals: They had to be reviewed and approved, and the NIH did not allow its scientists to consult for companies with whom they were also collaborating. Ultimately, though, the NIH leadership would argue that public-private collaborations could offer some advantages: The government could provide wider sample sizes for trials, as well as world-class expertise, whereas drug companies could offer deeper pockets. Ostensibly, the arrangement was a win-win.

Certainly, the pharmaceutical scene was heating up in those years. The Cognex that NIMH recommended to Jerry DeMoe in 1995 was quickly eclipsed when a newer, sexier drug from the same family arrived on the scene: donepezil, known by the trade name Aricept—which won FDA approval in 1996 and was sold in the United States by the pharmaceutical giant Pfizer.

Since Aricept's introduction, it has become one of the most widely prescribed drugs for Alzheimer's patients, ringing in more than $2 billion in annual sales before its patent protection expired in November 2010. It was what Karla's cousin Dawn was taking. (Cognex would be discontinued by its then manufacturer, the Japanese pharmaceutical company Shionogi, in 2012, after persistent reports about its damaging effects on the liver.) Perhaps the most poignant evidence of the public's desperation for an answer to Alzheimer's is contained in those sales figures, when even a drug that does not cure, and in fact only temporarily and partially staves off symptoms, is worth billions.

Looking to further the progress they'd just made with Aricept and to possibly develop new drugs, in late 1997 David Friedman, a PhD-trained neuroscientist for Pfizer, approached Trey Sunderland with a proposal: He wanted to create a three-way collaboration among NIMH, Pfizer, and the British company Oxford Glycosciences Limited to discover new, unknown biomarkers in Alzheimer's disease. Friedman had read some of Sunderland's published work and respected him because he seemed to recognize the importance of biomarkers—whether they signaled different stages of the disease or the potential for seemingly healthy people to develop Alzheimer's.

Sunderland cleared the agreement with his supervisor. It was the kind of public-private alliance the NIH encouraged, so long as scientists disclosed any potential conflicts of interest.

Under the agreement between Sunderland and Friedman, NIMH would provide scientific expertise, staffing, data analysis, and human tissue samples from Alzheimer's patients or at-risk subjects. The samples, in Friedman's opinion, were the least important item on that wish list; it was really Sunderland's expertise Pfizer was after. Pfizer had sought out Sunderland because they needed someone with "the experience, and knowledge, and access to samples that would make this project possible," Friedman said.

In total, Sunderland sent thirty-two hundred or so vials of plasma and spinal fluid to Pfizer for both the unknown biomarker study and the study of known biomarkers such as beta-amyloid and tau. The samples were worth approximately $6.4 million and had taken about fifteen years to collect. What made the Alzheimer's samples so valuable, according to Friedman, was that Sunderland knew the story behind the people who'd supplied them: What their family history was like. What their clinical symptoms were. He was the man who could connect the dots between any possible biomarkers and how the disease manifested itself in the person who carried them.

Pfizer wanted those vials to help develop drugs that would target the disease in its earliest stages. In April 2003, the storied *Journal of the American Medical Association* published some of their results in an article authored by Sunderland, members of his NIMH staff, and Pfizer employees. From David Friedman's point of view, it had been a very successful collaboration. But the company also used Sunderland to promote Aricept as an option to audiences of doctors during a televised NIH presentation in 2003.

Though he recommended the use of Aricept, Sunderland never told audiences he was being paid by its manufacturer; they only knew he was an NIH expert. In May 1998, he had negotiated his own private arrangement with Pfizer, under which he was paid $25,000 per year as a consultant, with separate payments of $2,500 per day for one-day meetings. All told, he earned roughly $500,000 from his five-year private arrangement, a payment that Pfizer considered "modest at best" compared to other, similar arrangements. Of that amount, Sunderland failed to disclose about $300,000 to his bosses. From 1999 to June 2004, he took the show on the road, speaking at US and international conferences while Pfizer paid him.

Aricept was a massive success by many standards. But there were some researchers—including one of Sunderland's own protégés, a young scientist named Susan Molchan—who believed Alzheimer's patients could possibly get the same benefits—a temporary slowdown in cognitive decline, for example—as these hugely profitable but expensive, privately controlled drugs provided from far more ordinary treatments. Dawn's daughters shared that skepticism about Aricept's benefits. And ultimately, it was the tissue samples that Sunderland had given to Pfizer, many of which had been collected by Molchan, that would spell his professional downfall.

Molchan was a graduate of the University of South Florida's medical school who went to work for Sunderland in 1987. Instead of moving on to a university after a few years at the NIMH, as most ambitious young scientists do, she committed herself to government service and to studying memory issues and ways to help Alzheimer's patients outside the traditional drug regimens.

One of Molchan's early studies examined the effects that stimulants had on memory; she wanted to know whether a drug like Ritalin or even a simple cup of coffee could confer benefits.

In 1993, Molchan embarked on a study that tested whether lithium—an element sometimes used to treat mania—might help prevent tau proteins from becoming toxic. Toward that end, she collected the spinal fluid of Alzheimer's patients, as well as healthy volunteers. By NIH standards, the lithium study was small: gathering samples from just twenty-five people. However, Molchan took more fluid than was typical, with the intent of storing some for future research, to which the patients each consented. Although Molchan was devoted to the study, it received little support from the institution at large and faded away without generating any concrete results.

Molchan began to think that her studies got less support because she was suggesting that caffeine or other mild stimulants could do the same job as Aricept for less money. During her five years at the FDA, she came to believe that clinical trials could be manipulated, allowing drugs to win approval even without proving much clinical benefit. She cited cholinesterase inhibitors such as Aricept as one example.

When Molchan arrived in 1987, she said Sunderland promised to support her for a tenure-track position. She and her husband had started a family that would grow to include three daughters. However, as it turned out, the atmosphere at NIMH wasn't welcoming for a young working mother. Very few women had tenure or were on a tenure track. Access to resources depended largely on office politics; Molchan and a colleague were both promised a research assistant, but neither person received one. If the women complained, they were sometimes told they had psychological disorders; according to Molchan, Sunderland informally diagnosed her as paranoid and depressed, while another woman was ordered to see a psychiatrist.

Perhaps predictably, Molchan's application for tenure-track positions went nowhere. Frustrated that younger male colleagues with less experience were getting professional advantages that she didn't have, she filed a sex discrimination claim that was dismissed. Finally, in 1996, she decided to move on and took a position with the FDA.

In 2001, a job opened up at the National Institute on Aging, one that rekindled Molchan's interest in Alzheimer's research. Three years later, in the fall of 2004, she was alerted to a lithium study requiring spinal fluid similar to the archive she'd left behind at NIMH. The colleagues who proposed the study suggested that Molchan retrieve the NIMH spinal fluid samples she had previously collected to use in the new project.

When she contacted Sunderland for access to the archive, however, he responded by sending a tiny amount of spinal fluid from eight Alzheimer's patients and two unaffected people.

"Now, I understand we didn't need a whole lot, but we would have liked a little more than a half of a cc," Molchan said. But when she asked what happened to the rest, Sunderland appeared to hedge. It had been a long time since she collected those samples, her former boss explained; moreover, some were lost through freezer failures.

True, at the time, the NIH's policies for tracking its human tissue samples were erratic. By 2006, the NIH was beginning to discuss ways it could centralize its tracking system through a computerized database and start requiring the transfer of all human samples to involve some form of written recordkeeping. But before then, how well samples were recorded depended on what lab they were in; some kept written records, others didn't. The only system-wide regulations the NIH had concerned hazardous biological materials. Otherwise, ordinary tissue samples such as Molchan's spinal fluid were largely untraceable. If a freezer malfunctioned, or samples spoiled some other way, nobody had to account for them. If they were stolen or used for purposes other than those to which the donors had agreed, nobody had any way of knowing.

However, for Molchan this atmosphere of carelessness didn't explain away Sunderland's vagueness about the vials, because Sunderland himself had always been scrupulous with the lab's samples, documenting how much was on hand and including additional details, such as whether the fluid had been collected at the beginning or end of the spinal tap. Every twenty-four hours, someone measured and recorded the temperature in the freezer. If it broke, or the temperature even fluctuated, alarms went off. Knowing what she knew about her former boss's habits, it seemed unlikely to Molchan that every backup system might fail, wasting so many people's sacrifices.

Ultimately, the fate of the missing samples would become the subject of a prolonged legal battle that would cast Pearson Sunderland III from the comfortable position of NIMH's chief of geriatric psychiatry to being a disgraced footnote in the history of Alzheimer's research.

When Molchan failed to get any answers from Sunderland or other channels at NIMH about the fate of her samples, she contacted the inspector general and the Department of Health and Human Services. Finally, in April 2005, she tried the Committee on Energy and Commerce, whose chairman, a Texas Republican named Joe Barton, had a mother who'd died from Alzheimer's. Her complaint got the politicians' attention, and the matter was brought before the House Subcomittee on Oversight and Investigations.

On June 13, 2006, she testified before the subcommittee about her concerns.

"As a doctor, my first obligation is to advocate for patients who put their trust in me," she said. Her testimony echoed the Noonans' demand that all of their biological samples be widely shared. "Some of these patients had contributed their time and bodies to a number of my research studies and others at the NIMH. These good people are always ready to help and work on Alzheimer's in any way my colleagues and I asked."

Some patients were fine with their tissues being used in any way the researchers saw fit. But not everyone felt that way. Submitting to a spinal tap—or agreeing to donate your brain postmortem—for the greater good of finding a cure was a noble cause, especially given the emotional investment and the fact that it subtracted from what precious time you had left to spend with loved ones while you still recognized them. Not everyone would have agreed to make those same sacrifices if they knew what Sunderland was doing. At the very least, it was a violation of trust. At worst, it could dissuade participants from research that could yield better results. Not everyone had the means or the energy to participate in multiple studies.

Molchan recounted the tale of the missing samples and her efforts to track them down. Congressman Barton asked her whether Sunderland ever told her to simply mind her own business.

He hadn't, Molchan said; but Insel, the NIMH director, pretty much had.

Paraphrasing what Insel had told her, she said, "Dr. Sunderland is occupied with plenty of other things right now. Please leave him alone."

For his part, when testifying in front of the committee, Insel did not hide his admiration for Sunderland, with whom he had once shared an office at NIMH, and admitted that he never asked Sunderland what happened to the missing tubes of spinal fluid.

Despite his place in Insel's good graces, Sunderland was nonetheless feeling pressure about the irregularities in his Pfizer agreement. On August 19, 2004—about the same time he met the DeMoes—the NIH's Office of Management Assessment interviewed him about the arrangement. He contended that he was not involved in any conflict of interest; at worst, he said, he had made some honest mistakes in his disclosure paperwork. The investigators were not convinced.

"Not disclosing over $500,000 [*sic*] in income was not an oversight or lapse in judgment but appears to be a deliberate decision not to comply with the rules, policies and procedures that are necessary to protect the NIH, its scientists and most importantly, its science," the Office of Management Assessment concluded.

The NIH recommended to the Commissioned Corps, which technically employed Sunderland, that he lose his job. In a cover note attached to the recommendation, Insel expressed his regret over the whole affair.

"I thought of Dr. Sunderland as one of the people who had made tremendous contributions to the agency," Insel said, but the NIH's Office of Management Assessment concluded if Sunderland had committed similar infractions as a civil servant, he would have been fired.

Even so, his golden-boy persona still carried some of its old luster. By the time his case reached the congressional subcommittee seven months later, Sunderland was still on the job—a fact that frustrated Barton, the congressman from Texas, as did the doctor's lack of cooperation with the congressional subcommittee staff charged with gathering information about his case.

"What little information that we have gotten, some of it appears to be misleading or intentionally inaccurate," Barton told Insel. "So we have a person who has, on the surface, suffered no repercussion—none—and you talk about a Camelot."

Nonetheless, at the hearing Insel argued that parts of Sunderland's relationship with Pfizer had been helpful. For example, Pfizer had tested one of the spinal fluid proteins with an antibody that was not widely available, saving the NIH a "tremendous" amount of money in the process.

Insel pointed out that the ethics case had been referred to the Department of Justice, and that he had done all he could do; he was Sunderland's supervisor, but he couldn't fire him. In fact, Insel seemed almost indignant, as if the unfinished state of affairs were somehow unfair to Sunderland.

"We made a referral. We are now seven months down the road, and this gentleman is still waiting to find out about his fate," he said. "The true north here has to be the question of separating out public role, official duty, from private [gain]. And if that wasn't on the table, then we are here to congratulate Dr. Sunderland on having done, I think, a really exciting scientific collaboration which . . . may actually bear some really important discoveries for families with Alzheimer's disease."

Barton was sympathetic, but only to a point. He would have far preferred that the hearing was called to highlight research breakthroughs. Instead, they were dealing with a prodigal son.

And for all the NIH's talk about toughening up its ethics standards, there really did not seem to have been any cultural change within the institution, Barton said; nobody seemed to be condemning Sunderland's actions. Barton concluded by saying such a lapse was "inexcusable."

On December 8, 2006—six months after the subcommittee hearings—Trey Sunderland pleaded guilty to a criminal conflict of interest as part of a plea bargain. Two weeks later, a federal judge in Baltimore spared him from serving any prison time, but ordered him to repay $300,000 to the government and sentenced him to two years of supervised probation. Sunderland said he planned to perform four hundred hours of community service at a retirement home for veterans in Washington, DC.

In one of the only public comments he would ever make about the case, Sunderland consulted some prepared notes and said that when he thought back on the events that led to his troubles, he could not explain what had happened.

"This process has humbled me in a way that I have never experienced before," he said. His voice quavering, he added, "This has been the most difficult thing I've ever had to do."

Despite her protests against his use of the samples, Molchan remained sympathetic to her former boss on a personal level. They shared several mutual friends. She wrote to the judge asking for clemency and drove up to Baltimore with her husband to witness the sentencing. He had once been Mr. Perfect, and she felt that he'd been a victim of his ambition. The other punishments were adequate.

Besides, she noted, other NIH researchers had done the same thing: "He's not as smart as them," she said. "Some of the guys got away with a lot more drug company money than he did, but they were smarter and didn't get caught."

After he lost his job with the NIH, Sunderland sustained further blows to his professional career. His medical license was revoked in Maryland in 2009 and in New York in 2011. In 2010, Sunderland applied to regain his license to practice in Maryland, but his application was denied. "He has not convinced the Board that he fully understands the deceitful characteristics of his conduct," said the denial.

Though much mention was made during the congressional subcommittee hearing of the sacrifices made by the volunteers who donated their tissue samples to the NIH, none of those volunteers was called to testify. One, who joined the research because his own father had died with Alzheimer's, did speak to a *Los Angeles Times* reporter after Sunderland's hearing, complaining that all the doctor got was a "slap on the wrist." Furthermore, he wanted the NIH to return the spinal fluid he had donated.

But others, even those who had worked closely with Sunderland, knew nothing of his troubles until after the fact.

Julie Noonan Lawson was deeply saddened by his conviction, as well as the implication that he enriched himself at his patients' expense. That wasn't the man with whom she had worked for four years, who had been so compassionate and careful in his treatment of the Noonan siblings.

"His ego didn't seem to be the thing that was leading him in his research. It was a cure that seemed to be leading him," Lawson said.

"He was just very human, very empathetic," she added. No other researchers had taken the Noonan family as seriously as Sunderland did, or offered much beyond the cognitive tests they had taken so many times that even the siblings stricken with Alzheimer's had memorized them. "I felt like he was a very good scientist, and he was pursuing areas that nobody else was pursuing."

The knowledge that he took money from Pfizer to hawk Aricept neither surprised nor offended her; in fact, he had been promoting Aricept when she first met him at an Alzheimer's conference. To Lawson, the NIH scientists were paid so little compared to their private-sector counterparts that such moonlighting almost seemed inevitable. (At the time of the investigation, a scientist in Sunderland's position would have been earning between $150,000 and $200,000 per year, according to Thomas Insel.)

But what bothered Lawson the most was that Trey Sunderland was no longer pursuing a treatment for Alzheimer's. In fact, the notion of that loss choked her up: the wasted years, the wasted opportunity.

When Sunderland lost his job, NIMH canceled his study. The information and samples he collected were warehoused, and over time, some of them were lost. Eventually, the study would resume, but with far less funding. Marilyn Albert, a neuroscientist from Johns Hopkins who had once praised Sunderland's work as being ahead of its time, finally took over four years later.

Albert was awarded $1.5 million per year for five years, but she was told specifically to limit the study to cognitive and clinical testing, eliminating the MRI scans and the collection of any additional spinal fluid. When the economy softened, her funding was cut by about $300,000.

In return, she received roughly fifty boxes of paper, a thousand electronic files, and five freezers full of specimens, along with several disks of MRI scans and a list of participants from their last point of contact. Of the roughly 350 volunteers that Sunderland had enrolled, 200 immediately said they would return to continue the study. With a little more outreach, including an agreement to visit people in their homes, Albert's group was able to get that number up to 300. They worked painstakingly to reassemble the abandoned study's information, which did not come with any kind of documentation or road map to help them.

The Noonans remain involved in several tendrils of Alzheimer's science, including Bill Klunk's study at the University of Pittsburgh, using the PiB compound he'd developed with Chet Mathis. He traced the disease through a series of clinical tests, comparing those results to the amyloid buildup he was able to see in brain scans. Eventually, he hoped his subjects might be able to test experimental treatments for clearing the amyloid.

Karla, who was devastated by the abrupt end to the Sunderland study that she thought would finally help her family, learned about Klunk's study from Julie and reached out to him.

One of the most poignant losses in the Sunderland affair was uncovered when Julie decided, on the advice of a doctor she'd consulted in Boston, to arrange for her family's brain tissue to be permanently housed in the same location for the sake of continuity.

But when she started inquiring about the whereabouts of the brains that her family had donated to the NIH, she made a horrifying discovery: Two of them were missing. They were the remains of her sisters Kathi and Maureen, who had died the same day on separate sides of the country.

Nobody could explain what had happened to them.

Marilyn Albert tried to help. She doubled back to the NIH, but the records were so incomplete that she wasn't able to get a definitive answer. One rumor said one of the brains went to a lab at the NIH, where scientists may have purged the freezer. But since the lab's records were incomplete, Albert couldn't confirm that. Another tip suggested that a brain had been sent to another institution, but when she tried to track it down, the investigator in charge of that research had retired, and apparently vanished in retirement, a similar situation to the one the DeMoes had encountered when they were unable to find June White.

Documentation was important to Marilyn Albert. Years earlier, she had been in a serious car accident that taught her a sobering lesson: One day, quite suddenly, you might not be around to explain your work to people. So for more than twenty years, she has been careful to record her research, creating a continuity that could survive an abrupt transition, offering a road map to a future scientist who might want to use that hard-won knowledge to better understand a problem.

When she spoke with people who participated in the study, Albert was always careful to express her gratitude that they were willing to donate their time and bodies in the search for a solution.

"That's the only way we're going to get the answers to these questions," she said.

Fifteen

FAVORITE SON

WHEN THE INITIAL shock of Dean's diagnosis had subsided, Karla tried hard not to worry about him. Unlike Brian and Doug, Dean had a wife to keep an eye on his affairs. Karla was willing to help, of course, but he was still several years away from the point where that might be necessary. She had more immediate problems to think about, particularly with Brian, whose deterioration was beginning to show.

Debbie Thompson, Brian's high-school girlfriend, had long since married and settled in Bismarck, where she worked as a medical technician. She'd seen and spoken with the DeMoes periodically through the years, although when she attended her thirtieth high-school reunion in 2004, she was surprised to see Gail there. But that was vintage Gail. Rather than being embarrassed, her children loved her eccentricities, as did Debbie, who remembered well the DeMoe sense of humor.

When Gail broke the news to Debbie about Brian's diagnosis, her stomach dropped. This man, still strong in middle age, who had teased and laughed and loved her in her youth, was nearly at the end of his life. She could hardly believe it.

At Gail's insistence, Debbie took one last turn on the dance floor with her first love.

She lingered late that night, reminiscing with Brian, and stayed in touch with him after that. On a subsequent visit back to Tioga, she stopped by the DeMoes' house, and Brian gave her a gift of earrings, a bracelet, and flowers.

"I still have them," Debbie said. "I'll keep them forever."

Brian was living in a trailer on the outskirts of Tioga when Karla and Gail began to think that maybe it was time to bring him closer to the center of town. With his former wife, Christy, long gone, they wanted to keep an eye on him as his memory and thought processing deteriorated; Alzheimer's patients often wander and go missing, or people take advantage of them.

They found him a little one-bedroom house just down the street from Doug's and a few blocks from Gail's. He moved in, bringing a stray cat that he'd adopted and named Missy, until he found out it was a male. Then he called it Mr. Missy. Brian's daughter, Kassie, found it ironic that her father, who had so hated cats throughout her childhood, was mellowing under the shadow of Alzheimer's to the point where he doted on his kitty. The first night he spent in his new house, he climbed onto the roof to rescue the cat, only to fall and break his leg in the process.

For a few years, Brian got by. In the aftermath of Sunderland's firing, Karla was determined to steer her family into another research venue. She hoped Bill Klunk's work at the University of Pittsburgh would fill that void. He ran similar tests to Sunderland's study, but he also took images of the family's brains to examine the amyloid content and compare it against the other findings.

It would prove to be a most fortuitous partnership. In Klunk, the DeMoe family finally found their champion: a man who cared about them as human beings as much as he cared about the scientific knowledge he could gain from their unusual genes. Gail saw Bill Klunk as a man who would watch over her family in ways that she couldn't; his entire staff treated them like old friends. One by one, the DeMoe siblings trekked to Pittsburgh each year, undergoing a battery of tests and brain scans so scientists could see how the disease was progressing both biologically, inside their brains, and clinically, in their behavior.

Although the study paid for their travel costs and provided a meal stipend and small honorarium, each study subject had to schedule at least three days off from work and convince a partner to accompany him or her—a spouse if the subject had one, or a close friend or family member who could objectively answer questions about the subject's day-to-day level of functioning. The study partner also had to schedule time off, and in some cases, they had to handle the rigors of traveling with an Alzheimer's patient. In more advanced cases of dementia, travel can be overwhelming and can trigger wandering.

Gail frequently traveled with her offspring, and Klunk and his staff came to know each of them intimately. When their children grew into adulthood, most of them joined the study, too, as did the Wisconsin branch of the family. Pat and Dawn were too far advanced in their disease to participate, but Dawn's sisters, Robin and Colleen, did enroll, as did her two grown daughters, Leah and Alayna. Karla tried to interest the Oklahoma cousins—Sharon and Jerry's two daughters—but initially, they decided to hold off.

Though the tests were grueling, Klunk and his group worked hard to make it as palatable as possible for everyone. The subjects were encouraged to sightsee, attend ball games, explore the city. When the workday was over, the DeMoes and the doctor often ate dinner together. Klunk did this with many of the patients in his study, with one exception: If they chose not to know their genetic status, he politely declined dinner, fearing he might accidentally tip someone off to what he knew.

To him, the DeMoes were more than just research subjects; they were people he admired for their courage and their selflessness, and their contributions were allowing science an unprecedented window into a disease that had baffled them for more than a century. He would come to know three generations of DeMoes; he would witness their losses, their triumphs, their deepest secrets and most paralyzing fears. He was humbled by their sacrifices and their trust in him, especially after their earlier disappointments; he would become their touchstone, a lifeline as they were battered by the relentlessness of the disease. Unlike anyone else they had met in the research world, he always made himself available, always returned their calls.

"There's only one family like the DeMoes," Klunk said. "They're the salt of the earth."

All told, Brian made it to Pittsburgh twice to participate in Klunk's study before the disease claimed too much of his brain to make travel possible. His small world grew smaller; he walked across the street from his little house to the Skol Bar, which was populated and run by his friends. They made sure nobody stole from him; it would have been easy to take his money.

At the local drive-in, family friend Kim Johnston worked behind the counter and waited on Brian when he showed up for lunch. Knowing that too many questions flustered him, she simply placed his standing order for a Philly steak wrap without asking and took the correct amount of money from Brian's billfold when he silently handed it over to her. Then he'd settle into a table across from her and wait for his food, saying, "Ayup," like he was relaxing after a particularly hard day at work.

Brian had always been generous to a fault; now he bought small trinkets to give his mother. They were all he could afford, but he still made the gesture to show her his affection. He walked to her house, often woefully underdressed for the harsh weather, loping along on his bum leg that had never quite healed right after the fall from the roof. On one memorable occasion, Kim Johnston taught him to ride a three-wheeled bike that someone had given the family, thinking it would help him get to Gail's more easily.

At five-foot-two, Kim struggled to hold the small bike while Brian moved the pedals, forgetting to coordinate each push.

"OK, now, now, now!" she yelled, as Brian started to get the hang of it.

"I can't!" he shouted.

"Can't never could," said Kim, borrowing one of Gail's favorite expressions, and Brian pedaled faster.

They wobbled down the street, past an oil company office, where Kim saw workers sticking their heads out the window, fascinated at the spectacle of a burly man in a trucker cap shakily pedaling a bike several sizes too small for him while a short woman cheered him on like an exultant mother. By the time they reached Gail's house, Gail was outside, laughing so hard she had to cross her legs to avoid wetting her pants.

"That was probably one of the last times that I thought part of him was still in there," Kim said.

Now that he was home all the time, he would spend hours playing the practical joke that had been a DeMoe family tradition for decades: canning cars. The trick never got old. As his disease progressed, it simply got more hilarious, the way repeated knock-knock jokes gain comedic momentum for small children. He canned cars so often that the people of Tioga came to expect it when driving past Gail's house; a few times, he forgot to let go of the fishing line and nearly lost a finger.

And sometimes, though he loved her, Brian was uncharacteristically cruel to his mother. When Brian's mood swung low, he unleashed his temper on Gail, much the way his father once had. Though she knew by now that the viciousness was Alzheimer's, it was impossible for Gail not to take the abuse personally.

Occasionally, she brought him with her when she went across town to volunteer at the nursing home attached to the Tioga Medical Center. Sometimes he slipped outside to smoke a cigarette while she was doing patients' nails. That worried her because she knew he was a wanderer, and the nursing home wasn't a locked facility. What would happen to him when he was no longer capable of living on his own?

The answer came soon enough, in February 2008. From the other side of the state, Karla had been monitoring her brother's living situation. She called the woman who tended the Skol Bar, and learned that Brian was sometimes emerging from the bar's restroom without pulling up his underwear. He was becoming incontinent, too—his cleaning lady reported that she was finding more accidents in his home bathroom. And Karla knew that the ugly incidents with Gail were escalating.

Since Brian was no longer married, his children were scattered, and Dean had been diagnosed with the mutation, Karla alone held power of attorney over his affairs. It was up to her to decide what to do next. After consulting with Brian's children and some of his friends, she made the first of what would be many difficult choices. She decided her big brother would enter a memory care facility—a step below a nursing home, but one that specialized in people with dementia and was locked to safeguard against wandering. It was in Minot, eighty miles away.

Karla told Lori and a few other family members and close friends about her plan, but made sure to keep the news from Gail. No matter how hard it was for her to relive the terror and pain of her final years living with Moe, Gail could never have agreed to send her favorite son away.

The morning they decided to take him, Lori and Karla—who were staying overnight at Gail's, ostensibly for an ordinary visit—met in the kitchen with Doug's daughter, Jennifer, as well as Kim Johnston. Gail woke to the sound of the gathering and wandered down the hall into the kitchen. When they told her what they were doing, she began to cry.

"Mom, I just know you would never want to let him go," Karla said gently, reassuring her that it was the only choice they could make. Hard as it was, Karla accepted full responsibility for that decision, hoping to spare her mother some of the pain. Let Brian hate her; she could recover. Gail never would.

"It was awful," Karla said. "I didn't want her to feel that way."

The decision to move Brian to a locked facility—however pleasant the accommodations were, however necessary the move—would be just the first of many such difficult turning points for Karla. She felt as though she was betraying the closeness her mother had tried to instill in her her entire life. With each decision, a limb of her family tree fell to her feet, her peace of mind eroding just a little more.

When the group drove by Brian's house, he was taking out the garbage.

"Hey, you wanna go to Minot?" Karla asked him. They were headed out on a shopping excursion, she explained.

"Hey, yeah, let's go!" Brian said, and hopped in.

While Karla, Lori, and Jamie's girlfriend took Brian to Target and then to a movie, Doug's daughter, Jennifer, and their friend Kim Johnston went to his house and collected his belongings. They brought them to the memory care center; Brian's son, Yancey, later got rid of the cat.

After the movie, Karla drove past the facility, pretending it was the first time she'd ever seen it.

"That's an Alzheimer's research place!" she told Brian. "We should stop and check it out."

Brian was amenable, so they stopped in and had lunch at the center. A staff member met with them and asked Brian whether he'd be interested in participating in some of their research.

Brian demurred. No, thanks, he said—he was already involved in Dr. Klunk's study at the University of Pittsburgh. He explained that he couldn't stay; his sisters were in town, visiting. He had to go back with them.

When the staff pressed him to change his mind, he became increasingly agitated. They offered him some medication; as a man now used to taking pills for his condition, he took it without asking questions.

"I don't want any problems here," the center's employee said quietly to Brian. "I'd sure like you to stay."

Finally, subdued by the medication, Brian agreed. "I won't give you any problems," he said.

But his compliance was short-lived. When he discovered that he wasn't free to leave, he became enraged at Karla for bringing him there. He demanded to speak with Dr. Klunk.

Karla conferred with Klunk, who agreed to talk with her brother, and suggested that she should also be conferenced in on the call. But when they spoke, Brian tore into Karla, demanding to know why she had dropped him off at the center. Though she thought she had mentally prepared herself to be the bad guy, he was so ferocious that Karla began sobbing; in self-defense, she hung up. Bill Klunk later called her back and apologized; even he hadn't anticipated how violently Brian would react.

As with Moe, Brian's mood swings eventually began to mellow out.

Within a couple of months, he accepted his new home. When family members came to take him out on day trips, he'd stay out for as long as he could tolerate, then ask to be taken back to the center and his nurses: "I've got to go back," he'd say. "The girls are worried about me."

Surprisingly, despite the bitter way their marriage had ended, Christy visited him there, too.

"He [was] still part of my life because of my children," Christy said. The years of anger and hostility didn't stop her from crying when she saw him. "It hurt to see him in that shape. I wish I could have remembered him in a different way."

Brian's comprehension of his living situation seemed to shift by the day. On one visit, a friend left the building to light a cigarette, turned around, and was startled to see that Brian had followed out right on her heels, undetected by the staff. Finally, she handed him a cigarette, and they stood outside the building, smoking and making small talk. They could have been back in Tioga on Gail's patio; he hardly seemed interested in escaping. They walked back into the building together. But within five minutes, Brian was complaining again about how he was in lockdown and not allowed to leave.

Brian's children were grown, but they did stay in touch with him, even if they weren't particularly close. Kassie had married an aspiring track star, Frankie Rose, whom she met in college; they moved to North Carolina, where she gave birth to a little girl, Brianna. So far, she'd wanted no part of an Alzheimer's diagnosis; she preferred instead to leave her fate unknown.

"I made a choice to live my life as normally as possible and leave it up to God," she said. But the question nagged at her every day. If she happened to repeat herself, she wondered: *Do I have it?*

Her brother, Yancey, was worried, too. He was always misplacing things.

"You know, you're just carrying that huge weight on your shoulders," Yancey said. "There's always little things—like you forget your car keys. I'm pretty forgetful about where I place things. Is this what it's like? Or am I just having an unorganized day?"

Perhaps influenced by his stepfather's military career, Yancey had joined the army when he was eighteen. He was Brian's only son, and he resembled his father closely.

"I'm already at peace with the fact that I probably do have it," he said. "I noticed a lot more of my dad in me than I see a DeMoe in her," he added, meaning his sister. "Like the way I sit; my dad used to sit the same way, with my hand up like that. He used to do it all the time. I already think I have it, so it doesn't make a difference if I find out or not. I'm not going to live my life any differently than this."

Of course, his physical appearance and his mannerisms had nothing to do with whether he had the disease. But families of Alzheimer's patients often look for clues to their future, even if they stop shy of obtaining a genetic test.

Yancey was preparing for deployment to Iraq when Brian, increasingly plagued by incontinence and his limited mobility, moved from the memory care facility to a nursing home in Minot. When Yancey came back, he had a few weeks' leave before heading to Alabama to attend flight school.

When he stopped to see his father, Yancey was rattled; Brian looked so frail, nothing like the man who used to put in hundred-hour weeks in the oil field. He was vacant, feeble; he needed a wheelchair to go outside.

Brian looked up at his son, now a man grown, a man who had been to the other side of the world and seen sights that Brian might only hear about; witnessing destruction that others only talked about; only to come full circle, back home, and see a much different, quieter devastation within the walls of a nondescript Midwestern nursing home.

"Hey," Brian said. "I know you."

When she visited Brian with her husband and daughter, Kassie wanted to believe that her father understood that he was holding his grandchild in his arms. He had pictures of Brianna in his room, and he did seem to know her. She could see it in his eyes.

When Kassie became pregnant with her second baby, she told her father before she told anybody else.

"I kind of knew my secret was safe with him," she laughed. "It's sad to say."

Now, telling him her joyful news, she was shocked when Brian seemed to reach a moment of clarity: "Well, you already have one," he said.

The baby girl, whom they named Kingston Dakota—after her father's roots in Jamaica, and her mother's in North Dakota—would be born in 2010. But Brian would not live to know her.

Part

THREE

You've been paid for by people who never even saw your face. Your mother's mother, your father's father. And so it behooves you to prepare yourself so you can pay for someone else yet to come. Whose name you'll never know. You just keep the good thing going.

—*Maya Angelou*

Sixteen

THE BAPTISTS AND THE TAUISTS

FOR DECADES, SQUABBLING within the Alzheimer's research field has persisted over whether amyloid or tau is the key driver in the disease.

The nicknames for the two camps, Baptists (plaques) and tauists (tangles), echo the near-religious adherence each has to their particular dogma, though there is much crosstalk between them, with each school of thought acknowledging that the other plays at least some role in the disease. It is a classic chicken-and-egg debate.

Since the 1980s, the Baptists have been in the majority. The human body produces the amyloid precursor protein (APP) in several organs, although not much is known about its function in a healthy person. Some studies suggest that in the brain, it helps direct the movement of nerve cells during early development. In a normally functioning body, enzymes cut the protein to create smaller fragments called peptides. One of those fragments is amyloid beta, which is thought to be involved in plasticity, or the ability of neurons to change and adapt over time.

In Alzheimer's patients, something goes awry in the process, the protein does not break down correctly, and the sticky pieces of beta-amyloid cling together like wads of chewing gum, along with pieces of dead neurons; they go on to form plaques, which interrupt brain function.

In the amyloid-versus-tau debate, Paul Aisen, head of the University of Southern California's Alzheimer's Therapeutic Research Institute and former director of the Alzheimer's Disease Cooperative Study, was an unapologetic Baptist.

In his eyes, the theory explained so much: Why a mutation that triggers the overproduction of amyloid results in Alzheimer's. Why 75 percent of people with Down syndrome—which is caused by an extra copy of chromosome 21, the same gene where amyloid precursors live—develop Alzheimer's.

He acknowledged that the disease includes other processes—such as the mangling of tau proteins inside neurons, disrupting the cell's ability to transport nutrients and ultimately killing it—and that those might be fine targets for other therapeutics. But he believed the best way to arrest the development of Alzheimer's before it destroyed memory was some combination of drugs targeting amyloid in various forms: Hit it in the bloodstream and hit it in the brain.

Simply put, "If your goal is to shut off the engine behind neurodegeneration, attack amyloid," Aisen said.

For years, he was fascinated with the earliest stages of Alzheimer's: What subtle brain changes foretold the devastation to come? How early could science detect those changes? And could doctors stave off future damage by treating those early symptoms before the memory began to crumble?

The majority of Alzheimer's research is dedicated to the premise that if scientists can somehow prevent beta-amyloid from clumping into plaques, they will successfully interrupt the disease.

But the tauists remain a vocal minority in the field, arguing that failed attempts to find an effective treatment can be explained by the fact that science is focusing on the wrong target—amyloid—when it should be focusing on tau.

The debate has turned bitter at times, with tauists arguing that Baptists are sucking up all the publicity—and ensuing research money—to the detriment of the greater, collective cause. Allen Roses of Duke University, who was famous for his 1992 discovery of apolipoprotein (ApoE) gene variants that heighten the risk of Alzheimer's in the general population, was dismissive of the amyloid hypothesis: "I can go into any cemetery and find a tombstone over a dead person. The tombstone didn't kill him."

Located on chromosome 19, ApoE is a gene that codes a protein that carries cholesterol in the blood. Since the early 1990s, Roses argued that ApoE's variations, or alleles—known as ApoE2, ApoE3, and ApoE4—hold the real key to Alzheimer's. The theory says E2 and E3 bind to tau and stabilize it so it doesn't twist into the tangles that pepper the brain in Alzheimer's disease. But the E4 allele does not cling to tau, and Roses said that's why people with that variant have a higher risk of developing the disease.

Baptists agree that ApoE4 is important, they just think it's for a different reason: In addition to not stabilizing tau, they say it also appears to promote the buildup of beta-amyloid. Tauists say the amyloid buildup is simply a side effect of the true problem, or the "tombstones" that Roses described.

One man who counted Roses as a friend was Eric Reiman, a tall, well-heeled man with salt-and-pepper hair and a gravelly voice that moves rapid-fire through the most complex of explanations. Reiman is now the executive director of the Banner Alzheimer's Institute, an important outpost in the world of Alzheimer's research housed in a freshly minted building a stone's throw from its parent medical complex in Phoenix.

Reiman frames the amyloid-versus-tau debate with a colorful analogy: "Think of Alzheimer's as: it starts with some kindling (amyloid) to ignite the fire (tau). By the time you have the fire, does it really matter if now you've addressed the kindling?"

Educated at Duke, Eric Reiman was on the faculty at Washington University when Bill Klunk trained there, before Klunk went on to the University of Pittsburgh and created the PiB brain-imaging agent. The two men stayed in touch as professional colleagues, and Reiman deeply admired

Klunk's dedication. Reiman was fascinated with the way different parts of the brain work in concert to orchestrate normal behavior, and conversely how they conspire to produce psychiatric disorders, but he did not work directly on Alzheimer's disease.

In 1988, he joined the local Alzheimer's Association chapter in Phoenix, Arizona, just to set an example for his children, who were then eight and three years old. At the time, he wanted to show them the importance of performing community service in an area separate from his professional life.

A few years later, in 1993, he read a story in the *Wall Street Journal* about the discovery of the ApoE connection by his friend, Roses, back at his alma mater, Duke. Reiman, whose expertise was in brain imaging, thought he might have stumbled across a way to hasten prevention trials in Alzheimer's patients.

Reiman theorized that if he began studying people at three different risk levels for Alzheimer's—low-risk, with no copies of the E4 allele Roses had noted; higher risk, with one copy; and highest risk, with two copies—he could start tracking their brains in middle age and see how they changed. In those days before Bill Klunk and Chet Mathis created PiB to image amyloid deposits, Reiman planned to look at other biomarkers that were considered predictors of the disease.

In Alzheimer's, it could be the way the brain metabolizes glucose, which doctors can view by injecting a person with radioactive glucose and visually tracing its path via brain scans. Glucose, the main sugar in the blood, is the primary fuel that the brain uses for energy. As far back as the 1980s, science has known that portions of an Alzheimer's-riddled brain do not metabolize glucose well—typically the parts of the brain responsible for memory retrieval, as well as planning for the future and daydreaming. Screening for diminished glucose metabolism could predict Alzheimer's well before symptoms appear. What was not known was what role this dysfunction played in the disease, although some researchers believed impaired glucose uptake helped to form the tau tangles. Another Alzheimer's biomarker was brain shrinkage, which doctors measured with an MRI.

Instead of studying fifty thousand people across twenty years, as a traditional prevention study might do, Reiman thought he could look at a couple of hundred E4 carriers over two years. His team would try a different prevention therapy every two years—an admittedly aggressive course of action—and report on the outcome. And while they were at it, Reiman thought perhaps they could also work toward developing a new standard of care for patients and their families. He was beginning to realize the magnitude of the disease's ripple effect, the way it ravaged even the strongest family bonds and crushed the health of overwhelmed relatives, just as it had contributed to Gail DeMoe's nervous breakdowns when Moe was still alive, and Sharon's struggles as she sought help for Jerry.

"Over half of family caregivers become clinically depressed; they're often frustrated, physically exhausted, feeling helpless and uncertain," he said. And the limitations of the health care system's reimbursement model—which often does not fully cover the cost of patient care—meant there was no money available to help them, he added.

"There's not even adequate reimbursement to do a full medical evaluation—certainly not enough incentive for people to learn how to help patients and family caregivers navigate the course of their illness and address the range of everyday questions they have, if they even knew to ask," Reiman said. "What are the financial issues, or the legal issues we need to address? Is there an elder law attorney that I might use? How do I get my loved one bathed? How do I get some sleep at night, or get a break without worrying?"

Pierre Tariot, who would become one of Reiman's closest collaborators, coined a phrase for the system's blasé attitude toward Alzheimer's: "Diagnose and adios."

Once again, Reiman knew the field could do better.

A few years later, the Banner Alzheimer's Institute was born, with Reiman serving as its executive director. Working as part of a consortium within Arizona, Reiman started by studying the effects of lowering cholesterol, publishing results in 2001. ApoE proteins transport cholesterol in the blood and central nervous system, and both the generation and clearance of beta-amyloid are regulated by cholesterol. But pharmaceutical companies, always conscious of their bottom line, were unimpressed.

Even if the consortium could prove that one of their drugs shifted a biomarker of Alzheimer's disease, such as glucose metabolism, they still wouldn't win FDA approval solely on that basis. The government had simply been burned too many times with other medical disorders. Recent medical history was littered with examples of treatments that offered spectacular results in changing the biomarkers of a disease but completely failed to produce any clinical benefit: There were cholesterol-lowering medications that didn't lower mortality rates; treatments that increased bone density but didn't decrease fractures in osteoporosis patients.

The FDA has a system in place that prioritizes its approval of new drugs to treat serious diseases for which current treatments are nonexistent or inadequate. Among these is its "accelerated approval" designation, established in 1992. These regulations say the FDA can approve drugs filling an unmet clinical need based on a surrogate end point, which is a marker—a physical sign or other measurement—that is believed to predict a clinical benefit. This is largely a judgment call by the FDA based on scientific support for the end point, and winning accelerated approval is a significant shortcut that can save valuable time in getting a drug to people who need it.

But achieving accelerated approval isn't easy. At a meeting with the FDA in the mid-2000s, a statistician stood up and made a chilling pronouncement. He said he would not consider a biomarker a viable way to predict a drug's clinical benefit unless he also saw eight clinically proven treatments in which the treatment's biomarker effects accurately predicted a benefit. In other words, in order to prove that treating a symptom was beneficial, he wanted to see eight versions of a cure.

A British colleague sitting next to Reiman leaned over and muttered, "Eric, I thought this was the greatest country on earth. If we had eight clinically proven treatments, what would we need a biomarker for?"

The Banner team knew they had to provide some kind of incentive for pharmaceutical companies to invest in their idea. They thought maybe the way to go would be to convince the FDA to fast-track its approval for drugs that showed promise in the shortened trials. Already, there were hints that the government would be open to that kind of flexibility as the public gradually became more aware of how serious the Alzheimer's crisis could become.

Shortly after the Banner Alzheimer's Institute was established, the Alzheimer's Study Group—a nonpartisan organization cochaired by US Senators Newt Gingrich and Bob Kerrey, and including such heavy hitters as Supreme Court Justice Sandra Day O'Connor—began to develop a plan for addressing the slow-moving storm of Alzheimer's.

During the group's early years, Gingrich traveled to Arizona and met with members of the consortium to discuss their idea of accelerated studies.

"Eric, how many prevention therapies do you think you're going to have to study before you find one that works?" Gingrich asked.

Reiman glanced around at his colleagues, and said, "Let's be conservative; let's say ten."

"How much do you think it would cost to study a prevention therapy?" Gingrich asked.

"Let's say . . . fifty million," said Reiman, although as it turned out, he was lowballing.

"So let's see if I've got this right," the senator said. "You're telling me we might have a way to be able to reduce the risk of Alzheimer's disease with a five-hundred-million-dollar investment to study ten treatments for a disorder we know, with a certainty, is going to cost one-point-three trillion dollars a year. Who in his right mind would study just one at a time?"

Reiman smiled his broad, eye-crinkling smile.

"*Exactly*," he answered.

As he wrestled with the catch-22 of biomarkers and clinically proven end points, Reiman formed a powerful partnership with Pierre Tariot, the son of a World War II French refugee who had established an enviable reputation in clinical trials for Alzheimer's patients.

At the University of Rochester, Tariot built an impressively large clinical trial program specializing in Alzheimer's disease and prevention. He was also a remarkably compassionate and beloved doctor, and when one of his patients died, her widower offered Tariot money to create a program in honor of his late wife. When the university declined to create the program, Tariot teamed up with Reiman in Arizona.

"I was considered a person with a lot of ideas and vision," Tariot reflected. "I think Eric is the only person I've encountered whose vision is grander and more inspiring.

"I said, 'Let's do this.' "

Seventeen

EXPELLED FROM EDEN

THE INTERNAL STRUGGLES that Lori McIntyre had been describing to her friends and family for two years were finally becoming uncomfortably apparent to the outside world. Where once she was able to make herself at home in a boxcar, she now became hopelessly disoriented with her frequent relocations as a railroad wife. She could no longer balance the checkbook or work a cash register.

The time had come, as she had known it would, to answer the question that had been lingering in the back of her mind for years: Had the disease marked her, as it had her brothers and father? Unlike Dean, she held no illusions about the answer; she was practical, focused, and ready to face down whatever came next.

Bill Klunk paid for Lori and for Doug's twenty-one-year-old daughter, Jennifer—who had relocated to Tioga to be with her father—to undergo genetic testing there as part of his study. They went to see John Martsolf, the same doctor from the University of North Dakota who had met with the extended family in 2004. When the results were ready, Martsolf said he wanted to discuss them in person, so everyone gathered in Grand Forks, North Dakota, for Easter 2006.

In true DeMoe fashion, the meeting was an extended-family affair: Steve accompanied Lori, and Jennifer brought her boyfriend, both of her parents, both of her grandmothers, and Doug's then girlfriend. Karla and Dean tagged along for added support.

Jennifer DeMoe was more frightened than she had ever been in her life.

Though the decision to know her results was terrifying, she made up her mind that if she tested positive, she would not have children, for fear of passing along the mutation. Since she had always wanted kids, she felt that she needed to know whether that was a dream she would have to abandon.

Unlike their parents, who had been blissfully ignorant of the mutation when they'd started their families, the next generation of DeMoes would face this added dilemma: whether to have children. It was a deeply personal decision, but one for which the outside world—even other family members—would judge them mercilessly.

Jerry DeMoe's two daughters, Sherry and Sheryl, were already married and raising children. The sisters were close, despite their ten-year age gap; even so, Alzheimer's was one topic they rarely discussed. Their mother, Sharon, worried privately that Sherry, her older daughter—who was approaching forty—was becoming absentminded. Sheryl, who was younger, wanted to get tested, so she would finally know her status. She just wasn't sure when she would be emotionally ready to do it.

The Wisconsin branch of the family was particularly split over the issue of children. When Dawn's daughter, Leah, got married and became pregnant, her grandfather Rob and aunt bluntly told her she was making a mistake, since she didn't know whether she carried the mutation.

It wasn't as though Leah was a stranger to the disease. Both her mother and grandmother were in nursing homes. In fact, all three of Pat and Rob's daughters—Dawn, Robin, and Colleen—had now been diagnosed with Alzheimer's—and the cure that Dawn had once expected still hadn't materialized. Dawn's sisters each had children but were now single mothers, further isolated by their rural location and the family's acrimony.

Just knowing whether the next generation carried the gene—apart from the question of whether to have children—was becoming a point of heated debate. Deb DeMoe, for example, begged McKenna and Tyler to stay unaware. Having felt blindsided by Dean's diagnosis, she wanted them to forget about the possibility of inheriting the mutation and live their lives as normally as possible.

Leah's husband had grown up in a family with handicapped siblings. He didn't see the value of Leah burdening herself with the disease. To him, all life had worth, and he would love her no matter what. Alzheimer's had already impacted her relationship with her mother; now that she had finally found some joy in her life, what was the sense of potentially throwing it all away? Her father, a Type I diabetic since childhood, agreed. It was important to focus on life, not worry about death.

Leah said if she had a positive test, she'd simply keep it to herself.

"If I got my results, and somebody told me yes, I wouldn't be like, 'Hey, I've got Alzheimer's!'" Leah said. In fact, she told very few people about her family's history; it was a form of self-preservation.

"I guess I don't want it to be five years down the road, and you do something weird, and they're like: 'Hmmmmm,'" she said.

But Jennifer was convinced that she needed to know. Although her mother, Lola, was remarried and living in Nebraska with her new family, she agreed to come with Jennifer to learn the results. When Lola saw Doug, she embraced him and said, "I'll be here for you."

Dr. Martsolf first asked the group how they'd like to receive the news: all together, or in private? They voted to learn all at once. The tension was palpable; so much was riding on those results. Karla could hardly stand to hear what came next.

Dr. Martsolf read Jennifer's results first: She did not have the mutation. The walls exploded as everyone screamed with joy, and Lola began bawling. Only Doug was silent, seemingly as oblivious to the good news as he had been to his own misfortune.

In the ensuing happy confusion, most of the family left the room. Karla stayed behind with Lori and Steve.

To this smaller group, Martsolf delivered the mortal blow: Lori had tested positive. Despite all the advance warning, Karla was devastated. Her sister had been right. Just as when they'd been growing up, Lori knew her own mind better than anyone else did. The sudden high from Jennifer's news plunged sharply into despair. Lori lived so far away, and Karla wanted desperately now to protect her, to take care of her, in a way she had never known before.

But Steve and Lori were quiet, impassive—much the way Dean had been, but for a different reason. He had been trying to protect his wife by showing her he wasn't afraid; Lori had been expecting bad news. The doctor asked Lori why she didn't react more dramatically.

"Because in my heart, I already knew I had it," she answered simply, and at that point, Karla could see how deeply despondent Lori actually was, no matter how carefully she had prepared for this moment. "It wasn't a surprise."

It was an Easter that Karla would never forget. The sister who had always been her polar opposite, this wildly unconventional girl, was now bound to her by the inheritance that was taking over their lives. Their mutual need to escape their father's long shadow had always been the one thing they had most in common; ironically, their failure to escape was uniting them again.

After the meeting, the family decided to go out together for an awkward lunch that was equal parts celebration and mourning. Jennifer remembered how Steve and Lori lingered in their car outside the restaurant: They were calling their daughters to break the news. Though the McIntyre girls had noticed their mother's subtle decline, they had hoped for a better outcome. It was not to be. Jessica, Steve and Lori's oldest daughter, immediately made plans to find out if she had the mutation.

Afterward, Lori and Steve stayed with Karla and Matt in Fargo, making the best of their Easter. At one point that weekend, Karla, Gail, and Lori lay in bed and talked about what they had just learned.

"Are you afraid?" Karla asked her sister.

"Not really," Lori said. It was the truth. She didn't feel sorry for herself. What she most regretted was the idea that she might never know her grandchildren; she'd always wanted to be a grandmother.

The second wave of Alzheimer's preyed on Jamie's already heightened sense of anxiety. With each positive diagnosis, he worried more; the doctors said there was only a 50 percent chance of each kid inheriting the mutation, but now they were dropping like flies: Brian. Doug. Dean. Lori. At least three of the five kids in his father's generation—Moe, Pat, and Jerry—had it, as did all three of Pat's daughters. *Fifty percent chance my ass*, the DeMoes thought, even though the doctors kept telling them it was pure bad luck. But then again, Karla didn't have it, and they were already past that 50 percent mark; maybe Jamie would finally catch a break. He didn't feel any symptoms. Maybe, just maybe, he had dodged a bullet. In 2006, right after Lori's diagnosis, he decided to get his own blood drawn. His confidence was on an upswing; though he was saddened by his siblings' bad news, he was so much younger than them—young enough to still feel invincible.

Dean, the brother he'd always emulated most, sat Jamie down for a heart-to-heart talk. He offered one piece of advice that resonated: "Live your life. You have no control over what's going to happen to you."

Dean could not have realized, until it happened, the power of knowing his own fate. It was a choice he hadn't fully appreciated, but afterward, he deeply regretted the knowledge and the change it wrought on his life. He worked hard to ignore it, to be the same man he'd always been. But some of the fallout, like the disease itself, was beyond his reach.

He repeated, like a drumbeat, as much for himself as for Jamie: "Whatever you find out, you gotta be the same person. *You're still the same person.*"

Two months after Lori and Jennifer learned their results, their doctor in Grand Forks, John Martsolf, sat in front of Jamie.

It was strange; his innate pessimism notwithstanding, Jamie was feeling pretty good. Maybe he just *knew* he was OK, the way Lori had known she wasn't. He'd been to Dean and Deb's house for a cookout the day before, and Jamie assumed the same kind of confident, relaxed demeanor that Dean always projected.

When he met with the doctor, he brought a few people with him for support: Karla, his mother, and his current girlfriend.

His carefree façade evaporated in a single moment, when Martsolf confirmed that he'd tested positive for the faulty gene. He was just thirty-five years old. It was the final straw; Gail's sixth and last child, the fifth of those six to carry the mutation. Only Karla and her children had escaped.

"I really didn't know what to say then, when they actually sat in that room and told you that you had it," Jamie said. He was simply shocked.

His girlfriend crumbled. She had a young daughter from a previous relationship and wanted more children; now that would be impossible. Jamie was determined not to pass on the mutation. Unlike his brothers and sisters, who'd already had their kids when they found out they were carriers, he had the option of disinheriting the disease. He was adamant: No kids. The relationship ended.

Though Jamie had been stunned, Karla thought he took the news surprisingly well. Maybe he'd been a worrywart for so long, he'd inadvertently been preparing for this moment his entire life. Secretly, just as she'd been so certain that Dean did not have it, she'd guessed Jamie might, thinking back to his learning struggles in school.

But their mother was another story. Jamie had been her baby. For seventy-one-year-old Gail, each diagnosis was a reminder of loss so profound that some days, she could not get out of bed. It was wreaking havoc on her health, too: she began taking medication for her heart, and she developed tremors in her hands that were so pronounced that Karla took her to a neurologist, who prescribed propranolol. Gail suffered from fibromyalgia, a syndrome that includes chronic physical pain, fatigue, and mood swings. She tried very hard to find reasons to laugh, but there were days when she could no longer summon the courage.

Worried, Karla called her mother almost every day; Matt began to fear for his wife's mental health. She waited until he wasn't around so he wouldn't realize how much time she was spending on the phone. The medication caused Gail to slur her words sometimes to the point where Karla couldn't understand her. Karla arranged for social services to check in on her.

Even Bill Klunk, who spent every day of his professional life facing the harsh realities of Alzheimer's disease, found the DeMoes' situation exceptional.

"The generation of six were tremendously unlucky," Klunk said of Gail's children. "Things tend to regress toward the mean."

Jamie was alone after he and his girlfriend broke up, but he didn't have to struggle on his own for long. A few months later, he caught sight of Chelsey Determan across the room at a wedding reception and asked her to dance. It was a moment that would change both their lives.

Chelsey knew Jamie DeMoe by reputation; Tioga was a small town, after all, and both of them had grown up there, albeit fifteen years apart. Jamie's nephew—Brian's son, Yancey—was a few years ahead of Chelsey in high school.

A petite blue-eyed blonde who gave off a streetwise air despite her small-town upbringing, Chelsey was a straight shooter, unpretentious; and though she never gushed with the type of artificial sweetness that some women dab on like perfume, she had a core of genuine kindness at the center of her well-guarded heart.

When Jamie first asked her to dance, she was wary of their age difference. But maybe it was his rugged good looks; maybe it was his vulnerability that appealed to that soft spot she had. Whatever the catalyst, soon she was in deeper than she had ever expected.

"It's the DeMoe charm, that's the thing," Chelsey said, smiling ruefully.

They'd been dating less than a year when Chelsey found out she was pregnant. She knew, as did everyone in Tioga, that some of Jamie's family members had Alzheimer's disease. But she didn't know that they had acquired it through a genetic mutation, or how dominant it was. And she didn't know Jamie himself carried it.

Live your life. You have no control over what's going to happen to you.

When she broke the news about the pregnancy, Jamie broke the news about his diagnosis.

"I remember, he brought me a rose in a little jar," she said. "It was really hard for him, I think, to tell me. I think he thought I was going to leave him, or maybe I wouldn't want to have the baby."

The full impact of the news would take years to absorb, but Chelsey Determan's resolve was cast on that day, in that moment, and it never wavered.

"He said, 'Does it bother you?'

"And I said, 'I love you, and it doesn't really matter,'" she recalled, her eyes wet with tears.

The DeMoe family did not, initially, take Chelsey that seriously. Jamie had gone through a lot of girlfriends, so they had no idea when he would actually settle down. It embarrasses Karla, in particular, to think about how she treated twenty-one-year-old Chelsey, as though she were just another of Jamie's soon-to-be-exes, easily discarded and soon forgotten.

It was not in Chelsey's nature to fight with other people who loved Jamie; she would prove her dedication quietly, in a thousand small trials that would have sent another woman packing—for example, when Brian and Doug became incontinent from Alzheimer's disease, she didn't hesitate to help with the cleanup. She endured the small indignities.

When she and Jamie moved in together—like Brian and Doug, into a house not far from Gail's—she opened a day care in their house, earning some money while still being available to take care of her own baby.

Their daughter, Savannah, was born on October 16, 2007. Her timing was impeccable; she brought much-needed joy to her grandma Gail. Despite the mutation's looming threat, Savannah represented faith in the future, that the simple joys the rest of the world took for granted—a grandchild's innocent face—would live on, even if Gail's own children did not.

As Savannah grew, Chelsey took her to see her grandma Gail almost daily. And gradually, Gail found a reason to get out of bed.

• • •

Just two blocks from Gail's house, Jennifer had set up an apartment in her father's basement. She cooked for him sometimes and made sure he took his medication. She found a job in a group home, and she settled back into the familiar routine of Tioga. The family was beginning to reorder itself again.

But then the relative calm of Doug's post-diagnosis life took a hairpin turn in March 2008, when a large pipe struck him in the face on the job.

With his face cut and his eyes swollen shut, Doug briefly lost consciousness. A supervisor took him to the hospital in Tioga, where he was loaded into an ambulance and taken to Minot.

Jennifer was in Minot with a client from the group home when it happened; her grandma Gail called her, hysterical. All Gail could say, over and over, was: "He was so handsome."

A surgeon reconstructed his face with the help of thirteen plates and thirty-five screws to repair the damage wrought by the pipe, but from then on, he would suffer repeated bouts of reactive airway disease—an asthma-like condition in which the airways temporarily narrow—and pneumonia, sometimes serious enough to hospitalize him for days at a time in the harsh North Dakota winters. The next time he went to Pittsburgh for his research visit, his results showed his cognitive decline had worsened. The team there recommended that he leave his job.

Workaholic that he was, once Doug stopped being an oil man, he became much harder for Jennifer to live with. He couldn't get the hang of operating his electric shaver or his cell phone. Once, when he asked Jennifer to help him figure out the phone, he was completely naked, and as unaware of that fact as a toddler.

She worried about Doug; she was working two jobs now, one in an office and one as a bartender. Even though she left the group-home position so she could stay around Tioga to take care of him, she wasn't home much. Doug didn't know how to cook, even in a microwave, so he ate three meals a day with Gail, who sighed when she talked about their daily routine, like a young mother whose patience is tested by a preschooler. Brian was in the nursing home in Minot, while Dean was still living near Grand Forks, on the other side of the state, close to Karla.

The days were long for Doug and Gail, and she sought out things for him to do. "Sometimes I put him to work vacuuming," she said. Like Karla, she often felt alone. Pat died that same year, at sixty-three; it was as though Gail's life were circling back to the early days of her marriage, when Wanda died, leaving behind a disjointed family that could not put itself back together again.

Doug's childhood best friend, Gary Anderson, had gotten married and moved back to Tioga. Like Gail, Gary searched for ways to occupy his friend. Doug helped him move into a new house, with Gary careful to back him up while he carried things upstairs. Doug—who had once been such a graceful swimmer and an accomplished athlete—was struggling with his balance, a common problem for Alzheimer's patients. But Gary was subtle, not wanting to spoil the sense of normalcy. When Gary worked one of his construction jobs, he picked Doug up each day and brought him along. Doug sat in a chair, and Gary and the rest of the crew told him he was supervising.

Though Doug didn't drive much anymore and never left Tioga, he began rolling his truck through stop signs; the family worried that he might hurt somebody. Since Karla held power of attorney for her brother, she called a doctor, filled out some paperwork, and arranged to have Doug's license revoked. He was incensed, but she recognized that it was necessary.

Doug's crying tore at Gail, who suggested that perhaps Karla give Doug his license back; Karla refused. She was beginning to feel like the Grim Reaper; whenever she made the cross-state drive from Fargo to Tioga, she had the sense that everyone was waiting for the other shoe to drop. But Karla realized that she could be more objective about her siblings' decline than could Gail, who would do anything to smooth over a bad situation for them. Karla had to be their reality check, and she did not back down from that role, even when it meant upsetting the people she loved.

"It seems like I'm always the bearer of bad news. When I come, I think they get scared of what's going to happen, because they know," she said. "I think my mother's even getting scared of me."

But there was no doubt that she was effective. When she came home, things got done. She was getting more practiced at taking care of the ugly tasks, which relieved other family members of those difficult burdens. It made her feel useful, but more than that, she believed it was her moral obligation, a kind of penance for having escaped the disease.

After Lori's diagnosis, Karla was ready to swing into action again. She already handled the legal affairs of Brian, Jamie, and Doug, and she lived close enough to Deb and Dean that she was able to keep an eye on them, helping out if they needed her. Only Lori was far away. She called Steve.

He was in the car when she reached him on his cell phone. Karla cut right to the chase: "When are you going to divorce Lori?" she asked. "When you do, she can come back home to North Dakota, where we can take care of her."

Within the DeMoe family dynamics, Steve had long been an outsider. Though he and Lori attended family reunions and stayed in touch, and their daughters were close with their cousins, there was no changing the fact that he'd been the man who took Lori away, and a subtle resentment lingered between him and her siblings. Though he and Karla were cordial, their differences would become apparent in their struggle over what was best for the woman they both loved.

To his credit, he set aside whatever immediate angry response he might have had to Karla's suggestion. Instead, he told her: "The line's bad. I'll have to call you back when I get home."

Hours later, she called him.

"We never finished our conversation!" she said.

"Karla, as far as I'm concerned, we never had that conversation," Steve said. They never discussed it again, and remained on polite terms, but Karla regretted having crossed that line, even though she'd intended to help her sister. She had only wanted to be practical. It was chilling, how practical the disease could make you become.

Dean hated visiting Brian. When they drove back home from Tioga, Deb would often suggest that they stop in Minot; it was on the way. It was hard to see his oldest brother so helpless, especially when Dean thought of what it might predict for his own future. Worse, the family was beginning to suspect that the nursing home was not giving Brian the care he needed. He was sleeping so much now; when Deb visited, the staff would tell her Brian had just eaten, but he didn't seem as though he'd been awake long enough for that to be true.

During one visit, Dean and Deb walked into Brian's room and found him lying in his own feces. Furious and humiliated, Dean removed his brother's diaper and changed him.

Once they got home, he sat on the edge of the bed and said, "Deb, *you* try wiping your brother's ass!" before breaking into angry sobs.

Deb just held him, as she'd once held Tyler, and they sat on the end of the bed for an hour, talking.

Deb's father, ever the insurance man, offered some practical advice: Get coverage for when Dean's ready for a nursing home. It was the same advice Steve McIntyre took before Lori agreed to learn her diagnosis. But Dean wanted no part of that future. He didn't even want to take any medication for his symptoms. Brian took Aricept and Namenda, another drug that was supposed to temporarily alleviate symptoms. Neither he nor Doug seemed to benefit from the drugs, and they tore up their stomachs. *The hell with that*, Dean thought. He planned to meet his fate dosed with nothing more than cigarettes, Mountain Dew, and whiskey. Anything else was just lining the pockets of pharmaceutical companies, the same ones who had paid people like Trey Sunderland thousands of dollars to promote a drug that had made no appreciable difference for the DeMoes.

Nor did he want to wind up in a nursing home, lying in his own waste.

"Don't let me get like that," he told Deb. "When I get to that point, I'd rather not be around."

The entire conversation rattled Deb deeply. She couldn't stop thinking about it, so the next time they went to Pittsburgh, she brought it up to Bill Klunk, telling him what Dean had said. The whole direction of that conversation was making her uncomfortable. She was a Christian and didn't share Dean's fatalistic views.

Dean asked his closest friend, Monte Olson, for the same favor: *Before I get bad enough to go into a nursing home, kill me.* For Monte, who would do anything for a man he considered a brother, it was an impossible request—but it is a fairly common one among Alzheimer's patients, particularly those who have witnessed another family member's decline and degradation. In fact, it's the very reason tests predicting the likelihood of Alzheimer's are so rarely permitted outside of research studies; until a treatment is available, the knowledge holds too much lethal potential. Even some research subjects, such as the *paisa* in Colombia, are denied access to that knowledge.

But with science's increased ability to predict Alzheimer's—whether through the genetic testing of ApoE alleles or imaging amyloid with PiB—comes a new debate among medical ethicists. Is suicide a rational response to a positive diagnosis, especially when—as is the case for people without mutations—there is still a margin of error?

"The looming prospect of identity annihilation, to the extent that it exists, may give pre-demented persons further weighty reasons to commit preemptive suicide," wrote the associate editor of the *Journal of Medical Ethics*.

Such ethical questions extend to other forms of dementia, too. In 2015, Susan Williams, widow of actor Robin Williams, revealed that her husband's suicide the year before had been prompted by Lewy body dementia, a disease that can cause Alzheimer's-like symptoms but is associated with a different kind of protein.

Unlike other terminal diseases, such as cancer, dementia raises a specific legal dilemma: In states that permit assisted suicide for the terminally ill, the patient must demonstrate the ability to make rational decisions. Once an Alzheimer's patient reaches later stages, meeting that requirement becomes impossible.

To Deb DeMoe, Klunk suggested a therapeutic lie. "Promise him that you'll do as he asks," Klunk advised her, "give him the peace of mind. When the time comes to make good on your word, he'll never remember that he wanted you to do it. And in the end, the words you said today won't matter."

Dr. Klunk's advice would echo back to Deb in the coming years, as

Alzheimer's finally began to worm its way into her relationship with Dean. When Dean began to repeat himself, or forget details that she'd told him, Deb would get frustrated and remind him: *I already told you that several times.*

Klunk told her to stop. There was no scorekeeper in their house, keeping track of who was right. There was nothing to gain by reminding Dean of what he'd lost.

"You're going to have to be wrong a lot, Deb," he said.

And Gail, veteran of many such conversations, agreed.

"Gail said that was hard: the patience you need to be wrong," Deb said. "It's kind of like—and this sounds terrible—but it's almost like a slow suicide. You don't take your own life, but it's taken so rapidly from you."

In the waning months of 2010, Brian's decline seemed to accelerate. Nobody in the family lived close enough to Minot to monitor his care, but when they did check in on him, they were alarmed by what they saw. He always seemed to be asleep, his eyes shut and his mouth half-open, like a man who had dozed off in front of the TV. A woman who was in the home visiting her husband reported to the DeMoes that Brian seemed to have an infection on his arm. When family members called, the nursing home staff insisted they were treating him, but the family had no way to verify that.

Karla asked for hospice care, but the nursing home ignored her, saying he wasn't ready yet. She wanted to spare Brian the same misery in his final days that Moe had endured, but she didn't know how to fight the system.

About a week after she heard about the infected wound on Brian's arm, Karla answered another call from the nursing home: They wanted permission to bring Brian to a hospital to treat him for pneumonia. As a matter of fact, the nursing home employee told her, he might not survive.

Immediately, Karla dropped everything and went to Minot to see her brother. She was met by her best friend, Roxanne, and by Dean's elder daughter, Lindsey.

At the hospital, the doctor told her Brian should have been in hospice care, which infuriated Karla. He was placed on oxygen and given morphine, and the hospice promised to send someone to pick him up the next day.

But by the next morning, Brian's condition had worsened. The doctor offered them the option of taking him off oxygen, explaining that it would be the end; Gail had arrived, and the family consented.

She gathered her son into her arms for the final time: her dearest boy, who looked so much like his father. All the color drained out of Brian, and less than a minute later, he died. He was fifty-four years old, four years younger than his father had been when he died two decades earlier.

The death of her father prompted Kassie to begin rethinking her decision to remain ignorant of her status. Four months after Brian was buried, she joined her family on the annual research trek to Pittsburgh and allowed the nurses to draw her blood.

She told the family she had no intention of finding out the results, and that she was just getting tested to contribute to the research. It was a lie. She just didn't want to burden her grandma Gail if the news turned out to be bad.

Her husband, Frankie, didn't want her to get tested. He had watched his father-in-law's decline, and he knew the genetic risk Kassie carried—and the potential that she could pass the mutation along to their daughters. None of that diminished his love for her or scared him away, and he had seen the burden that knowledge of a positive diagnosis could bring by watching Doug, Dean, and Jamie.

But with motherhood, Kassie was increasingly worried about her children. So when she insisted on learning her status, Frankie prepared himself for the worst.

Time passed; nobody called. Finally, Kassie phoned the research coordinator and asked when she could set an appointment to learn the test results. The coordinator promised to have Dr. Klunk call her back to schedule a review.

Kassie waited some more. The uncertainty was killing her; should she read anything into how long it had taken them to get back to her? She tried to make peace with it, to believe that it was part of God's plan.

Then the phone rang; it was Klunk. She was expecting him to schedule an in-person visit, but in his excitement, he blurted out the news: She did not have the mutation. She was free, and so were her two little girls.

"We've cleared three people today," he said, and his joy was unmistakable.

When Frankie came home, she told him: "I have some good news and some bad news to tell you."

"OK . . . ," he said.

Kassie smiled her million-dollar smile.

"The good news is: I don't have Alzheimer's," she said. "The bad news is you have to live with me the rest of your life."

Eighteen

SAFE HAVENS

A YEAR AFTER Brian's death, a few of Dean's old friends from Tioga called him with a proposition: Had he ever given any thought to coming home and working in the oil fields?

Owing to a combination of technological advancement and the country's never-ending thirst for energy sources, the western half of North Dakota—which had been drilling oil since Gail and Moe were newlyweds—was undergoing a boom such as the state had never seen, and Tioga was at its epicenter.

For decades, oil companies had known this twenty-five-thousand-square-mile reservoir of oil existed under the Bakken Formation, the oil-rich spread of rock stretching under parts of North Dakota, Montana, Saskatchewan, and Manitoba. The problem was how to extract it in a cost-effective way. The solution was hydraulic fracturing, or fracking, a technique which forces a solution of water, sand, and chemicals into the rock at high pressures, cracking it and allowing the oil to escape through a well. As the technique was perfected, and oil prices rose, the stage was set for a black gold rush.

In 2001, less than 2 percent of North Dakota's oil came from the Bakken; in 2011, more than 80 percent of it did. Suddenly, North Dakota found itself compared to other powerhouse energy states, including Alaska, California, and Texas.

The boom could not have had better timing for Dean. It was nothing for a laborer to pull down $100,000 a year; for a guy like Dean, who was known for being a hard and meticulous worker, the opportunities were plentiful. And knowing that his working days were numbered, he wanted to do whatever he could to create financial stability for his family.

Returning to the oil fields meant being separated from Deb and the kids during the week. The idea made him terribly nervous, and that was a feeling to which he was not accustomed. But he was forty-nine years old, fighting a degenerative brain disease. As with any health crisis, particularly one affecting the family breadwinner, Alzheimer's disease can sap savings, causing financial problems from which families struggle to recover. By now, Tyler was enrolled at the University of North Dakota, and McKenna was in high school. Lindsey was finished with graduate school. Dean's father had died at fifty-eight, his brother at fifty-four, and both were forced to stop working when they were close to Dean's age. He took the job.

Because of his reputation for leadership, the oil company made him a tool pusher, a position that involved overseeing the rigs and supervising a crew. Oil hands liked to joke about how little tool pushers did; in the fields, it was a relatively cushy job. But Dean could never be content with such a passive role. Instead, he set about learning the job of every man on his crew, so he would know whether they were working as hard as they could. Once he learned it, he'd often embarrass a slacker by jumping in and doing the task himself. Though he hadn't worked in the oil fields since 1989, opting instead to work for an asbestos removal company, he became the best tool pusher some of the company's men had ever seen.

With Doug still at his house with Jennifer, Gail was living on her own, so Dean moved into his old room in the basement. Karla was relieved that he'd be available to keep an eye on things; despite the occasional forgetfulness that Deb and his children noticed, he was still very much his old self.

Every morning, before he drove to work, he left two cigarettes for Gail on the kitchen table. On his days off, he drove four and a half hours home to see his family; with Tyler, he'd go out to the bar for a beer. With McKenna, he cheered her on at her volleyball and basketball games.

Having Dean around offered seventy-six-year-old Gail some security, which she needed for the first time in the fifty or so years she had lived in the town. Apart from her failing health, Tioga suddenly found itself overrun with roughnecks coming to seek their fortune from all corners of the United States, and with them came a startling side effect: From 2005 to 2011, violent crime in the Williston Basin, which includes Tioga, increased 121 percent.

In the old days, Gail never minded when Moe's buddies wandered down the street from the bars and let themselves in to raid her refrigerator. But those days were long gone; now, when men came begging, Dean chased them off.

One fall morning, Jamie and Rikki Rice, his best friend, went out for breakfast. It was only 7:30 a.m., and the air was still chilly. As they headed to the café uptown, talking and laughing, they noticed some guy in the street. His hair was wet, as though he'd just stepped out of the shower, and he was holding up his pants and trying to run. In Tioga, strange sights were not so unusual anymore, with the oil boom attracting every misfit west of the Mississippi.

"Look at this goofball," Rikki said, and they laughed.

And then, just as abruptly, their laughter stopped, because they realized they were looking at Doug.

"Holy fuck," said Jamie. His eyes filled with tears.

They pulled up and offered Doug a ride; he was trying to make his way to Gail's. He refused their offer. Not knowing what else to do, they continued on to the café, and pretended the entire episode had never happened.

"That was a rude awakening for me," Rikki said. In a few months, the weather could easily dip to thirty below, and if Doug attempted the same journey and got lost, he could freeze to death. Even in large cities,

Alzheimer's patients frequently wandered off from their families, sometimes with tragic consequences. The Alzheimer's Association reports that more than 60 percent of dementia patients wander, and if they are not found within twenty-four hours, up to half of them die or are seriously injured.

Wandering is a hallmark of Alzheimer's because of the disorientation the disease causes. Unable to recognize their surroundings, patients might believe they have to get home when they're already home. Adding to the problem is the fact that Alzheimer's patients sometimes think they are in a different time in their lives, because the disease destroys the most recent memories first, so they could begin looking for a landmark from years past that only they would know.

It would not be the last time Doug was found staggering through the streets. He was becoming another of Tioga's many odd town fixtures. Complicating the deterioration of his mental state were the respiratory troubles that had plagued him since the industrial pipe smashed his face in 2008. With Jennifer working so much, he was often alone, so he kept trying to get to the only safe haven he knew: his mother's house.

The following July, Jamie's girlfriend, Chelsey Determan, was cruising along in her big SUV when she spotted Doug by the side of the road, stumbling, struggling, his face flushed, unable to breathe. It was sweltering, with temperatures topping a hundred degrees.

Shocked, she pulled up beside him.

"Get in," she said firmly. She was young enough to be his daughter, but Doug obeyed. She recognized how critical the situation was and drove straight to the hospital, calling Gail en route so she could meet them there.

It turned out to be another bout of reactive airway disease. Doug was lucky that Chelsey found him; his condition was bad enough to keep him in the hospital for six days. Worried that the outcome could have been much worse, Karla contacted the nursing home adjacent to the hospital to see if he could simply transition into their care.

At first, the home's staff demurred. Some of them had been friends with the DeMoe boys and shared their misspent youth; they knew how wild they were, particularly Doug. Everyone had known Brian, of course, and they'd heard how combative he'd been when he was locked up. They weren't really equipped to handle a working-aged man of Doug's size and strength; if he fought them, they feared they wouldn't be able to keep him—or the other residents—safe.

But Karla wouldn't accept a refusal. Brian's experiences in the Minot nursing home were still fresh in her mind. She could not, *would not*, allow that to happen again. She begged, cajoled, pleaded, played on their sympathies. She even persuaded the local Alzheimer's Association chapter to intervene and offer additional training.

"I don't know how Mom would deal with it if he wasn't in town," Karla said. "It would kill her to not be able to see him."

Everyone in town knew Gail; she was still the most popular woman in Tioga, a regular churchgoer, the belle of the senior citizens' center. When Dean took her out to the bar, all the locals stopped to buy her a drink. Nobody wanted to compound her losses.

Finally, the home's administrators caved. Since Doug was used to going to the hospital so frequently, Karla asked his doctor to tell him he needed to return for more recovery time. The hospital's beds were full, they lied—a plausible lie, given the frequency of industrial accidents occurring in the oil fields. They explained to Doug that he would be taking up residence in long-term care while he waited for a bed to open up.

Doug was convinced. Meekly, he moved into the home. At first, he thought he was in a hospital; but in time, he began to refer to the facility as "my apartment." He was, by far, the youngest person in the hallways; but Doug never seemed to notice. Gail or one of his other family members came to visit him and take him on outings nearly every day.

Just as Brian had done in his early years at a care facility, when Doug had had enough of his day trips, he announced that he needed to get back to his apartment, and his family complied; they were just glad he wasn't fighting with them the way Brian initially had.

Gary Anderson came to visit, and Doug smiled when he saw him, but it pained Gary to see his wild, reckless childhood buddy so changed. He didn't stop by as often as he felt he should. Among the many invisible side effects of Alzheimer's is the collateral damage it can do to once-solid friendships. People who are on the outside, looking in, often pull away, having no point of reference to guide their reactions to the disease. They avoid the uncomfortable situation, but in doing so, they are unwittingly isolating both the patient and the patient's family at a time when they need support the most.

Gary took some comfort in the knowledge that Chelsey frequently brought Savannah to see her uncle Doug. His friend had always been a sucker for little children.

Nineteen

A BIG IF

ERIC REIMAN'S NEW research partner, Pierre Tariot, arrived in Phoenix to an empty building in 2006, charged with a weighty task: to design a clinical trial to test experimental Alzheimer's drugs in patients who did not yet exhibit symptoms of the disease. Reiman, the brain-imaging expert who now headed the Banner Alzheimer's Institute, had persuaded Tariot, whose own expertise was in designing trials, to leave his post in upstate New York to tackle the disease as aggressively as possible.

Their conundrum, which had faced them since Reiman first started talking about clinical trials, remained: How could they achieve a result significant enough to win support from the government and capture the attention of the pharmaceutical industry without spending twenty years on the process? And could they work tests on biomarkers—such as telltale proteins in spinal fluid, or a slowdown in the brain's metabolism of glucose—into the mix, so they'd know whether a shift in the composition of the biomarkers predicted a drug's success or failure?

Tariot and Reiman worked so hard designing the study that Tariot began taking off whole days where he didn't answer the phone or respond to email; he just read, thought, wrote, and talked to the world's experts.

"That alone probably was a critical element for us to make progress," he said. "Otherwise, you spend your days reacting to the crisis du jour rather than saying, 'Now, wait a minute: How are we going to do this?' "

The two doctors started by asking the best specialists they could find in other fields. With their help, Tariot and Reiman identified combinations of tests that were sensitive enough to detect and track what cognitive decline looks like in people years before they showed any outward signs of dementia.

But where would they find people on whom they could use the tests? There was one group that could not be ignored: people who carried the autosomal dominant genetic mutation.

Reiman's and Tariot's cognitive tests were designed to work on patients as young as thirty. Though these patients functioned and tested normally, the results would tease out subtle shifts in brain function. But this realization brought a cascade of additional questions: When did those changes begin? Did the subjects already have evidence of brain damage that could be seen on an MRI? Did they have amyloid deposits? Were there abnormalities in their spinal fluid? Which would change first, the biomarkers or the clinical measures—and would they have any relationship to each other? What factors influenced the age at which the disease began to show?

"In our field, there are no answers to those questions," said Tariot. But researchers in other diseases—notably HIV and cancer—had already pioneered ways to find the information the Alzheimer's field sought. Among them was Don Berry, the lead statistician for the MD Anderson Cancer Center at the University of Texas. Though Berry's normal consulting fee would have priced him out of their budget, after hearing the Banner Institute's sales pitch, he spoke to Reiman and Tariot's team for free.

By January 2011, the team was ready to present its plan to government regulators. At one of its public meetings, representatives from the FDA and its counterpart, the European Medicines Agency, agreed to attend.

Though the meeting was not formal or legally binding, both agencies indicated that Banner was moving in the right direction, and that regulators would be flexible in their approval because they recognized that the Alzheimer's problem was reaching critical mass. The first baby boomers turned sixty-five that year, and the Alzheimer's Association was publicly predicting that one in eight of them would develop the disease—or roughly 10 million people. Of those who lived to eighty-five, nearly half would get it. Time was running out.

Prior to the meeting, Banner already had interest from pharmaceutical companies, which posed an entirely different problem. As Bill Klunk and Chet Mathis had found when they were developing PiB, companies that were investing considerable amounts of money in a treatment wanted some control over its development, even if the scientists who invented the treatment didn't agree with their decisions.

"If we proposed a study like that, where on earth are we going to find enough people to actually perform a sufficient test of the treatment, if there are only five hundred kindreds [with the mutation] around the world?" Reiman asked.

It was then that a couple of his colleagues reminded him of two names he'd heard before: Ken Kosik, who'd traded Harvard for a corner office overlooking the Pacific Ocean at the University of California, Santa Barbara, and Francisco Lopera, who had been buried for years in South America, mapping the constellation of an unthinkably large Colombian family who carried the PS1 mutation for Alzheimer's—the same one the DeMoes carried. Three years before the 2011 meeting with regulators, Reiman made the phone call.

In Colombia, Lopera was in the middle of discussions with Swiss pharmaceutical giant Novartis about a collaboration involving his *paisa* families. After decades of frustration, it finally seemed as though he was going to be able to deliver some hope to people who had lived for generations with none. But in the midst of those talks, Kosik received the phone call from Reiman.

Kosik persuaded Lopera to meet with Reiman and Tariot at the Banner Institute and hear what they had to say.

Immediately, the two groups recognized in each other a common bond, one that Lopera hadn't sensed in his discussions with Novartis.

"I saw that for me, it was easier to work with academic people than with pharmaceutical people," Lopera said. "It was a big responsibility to take the decision about what kind of treatment was best for my families."

The Colombian team wanted to work in what is known as a pre-competitive atmosphere, which would allow them to remain impartial when they selected a drug treatment. They didn't want that decision being driven by business factors, only academic goals, avoiding the same kind of conflict of interest for which Trey Sunderland had been rebuked. They wanted to pursue the treatments that made the most scientific sense, not the ones for which they'd been paid. In essence, they wanted the "Camelot" of pure, impartial science that the NIH had purported to be.

Of course, academia has its own problem: the currency of credit. Scientific research is littered with examples of professorial catfights over who was first on the scene in an important discovery, and with good reason: Being first is tantamount to being best. When Jonas Salk announced that he'd created a polio vaccine, he became an instant demigod, feted and honored for the rest of his life; when a pilot announced Salk's presence on an airplane, other passengers reportedly burst into spontaneous applause.

But years before him, a virologist named Hilary Koprowski mixed his own vaccine in a kitchen blender and drank it, inoculating himself against polio in 1948, seven years before Salk introduced the injectable killed-virus version that made him a legend. When Koprowski died, the *New York Times* quoted a historian who labeled him "the forgotten man."

In Alzheimer's disease, Dmitry Goldgaber's success in locating the APP mutation on chromosome 21 bested other teams by the narrowest of margins; working with Family N and other subjects, Peter St. George-Hyslop reached a photo finish in identifying the PS1 mutation. Achievement begets admiration, but it also invites rivalry, and Kosik was wary of setting the Colombian families up to become pawns in an institutional battlefield.

But the Banner doctors were open and generous; immediately, the teams sensed a common purpose, and the bargain was struck. "This is the first time anybody has offered something that could actually help our families," Lopera said. They began talking in 2008, and they formally established the collaboration in 2010.

The Banner team's arrival in Colombia would prove to be a seminal experience for both the doctors and the *paisa* families. Like most people, Reiman was somewhat apprehensive about the visit, as Kosik recalls. He was, after all, planning to launch a clinical trial in a country famous for kidnappings and guerrilla warfare. The team met with some of the families, unannounced, in a little schoolhouse on the outskirts of Angostura.

Tariot was struck by the warmth of the people and the terrible beauty of the landscape. Here, superstition dwelled side by side with traditional medicine; *curanderos*—traditional Latin American shamans—practiced alongside doctors. In Colombia, these healers also sometimes pull double duty as witches.

"Sometimes the job is to cure the person, and sometimes it's to curse them," Kosik explained.

Even to the most sophisticated doctors, it was a novel experience to search for the answers to their medical mystery in the far corners of the world.

Sitting with a group of *paisa*, Tariot told a fifty-four-year-old man that his sons were participating in research that they hoped would work toward a better treatment: "We think your sons are heroes," he said.

Tariot and Reiman would never cease to be in awe of people who, even in the lowest point of their lives, chose to offer what little time they had left on earth to a field that had largely ignored them, knowing their contributions might only benefit people they would never meet.

"You cannot meet these families and not be transformed," said Reiman. "Cannot happen. Pierre and I have met close to a thousand family members in Medellín. And you go into their home, and you see several family members affected at the same time."

His voice rose as he described what he saw: "Struck in the prime of their life, not able to work, and their children gave up work to care for them. And they've been living with it one generation after another. It's almost treated like a curse. To go in there and to talk, it's both a privilege and a keener sense of responsibility."

Many of them were shamed or frightened by Alzheimer's. It was difficult to parse out the medical knowledge these doctors brought from the superstitions of the *curanderos*. But still, they agreed to participate.

Reiman polled the family members, asking: "If we could provide one service, what would it be?"

The answer told him everything he needed to know about the circumstances in which they were living: adult diapers.

The thought that his team could help empower people who had lived for generations with a nameless, hated disease was exhilarating. It was science at its best. At the same time, he had to remind these people that what they were agreeing to do was experimental. There were no guarantees that whatever drug they chose would work, or even be safe or tolerable.

"We're going to learn a tremendous amount from this single mutation," Reiman said. "But those are all big ifs: A big if whether the treatment will work; all sorts of things can go wrong in clinical trials. And a big if [whether] one or more of these clinical biomarkers predict a clinical outcome." In other words, if a drug successfully reduced amyloid or tau in spinal fluid, for example, it might predict that years down the road, Alzheimer's symptoms such as forgetfulness or aggression would be delayed or prevented.

Scans had illustrated brain changes in mutation carriers decades before their symptoms appeared, convincing Reiman that treatment had to begin even earlier than many in the field thought—before the age of twenty-eight. At a minimum, the Colombia trial—which targeted only amyloid, since tau prevention treatments were not yet available for human trials—should be the definitive test of the amyloid theory, saying once and for all who had it right: the Baptists or the tauists.

But clinical trials are expensive, and medical research is a results-driven field.

"By the way, the field's not going to give us that many more shots on goal," Reiman said. Drug companies, feeling pressure from their investors, would not keep sinking money into expensive failures. And many researchers depended on such funding to do their work, and on successful research to keep their jobs. "So we'd better get it right."

Twenty

EVERYONE SEES THE POWER

WHEN BILL KLUNK was a graduate student at the Washington University School of Medicine in St. Louis, he first crossed paths with John Morris, who was completing a postdoctoral fellowship. Because both men were interested in the science of brain disorders, and specifically Alzheimer's disease, they continued to communicate as professional colleagues. Thirty years later, that relationship would lead the DeMoe family to their first real shot at a meaningful treatment.

Like Reiman and Tariot, Morris believed that Alzheimer's formed much earlier than anyone had ever suspected, and that identifying the right time to administer a drug was the key to preventing—or at least delaying—the disease.

In 2008, in an effort that paralleled what was happening at Banner, Morris began taking his first steps toward early prevention. From Washington University, where he was now a professor of neurology and head of its Alzheimer's Disease Research Center, he launched an international study of people who carried the known Alzheimer's mutations. Klunk, who headed the sister center at the University of Pittsburgh, contributed his patients—with their permission—to Morris's larger effort.

Morris called his study the Dominantly Inherited Alzheimer's Network, or DIAN. He sought to enroll four hundred people who came from families with the mutation, though they didn't have to be carriers to participate—siblings like Karla, who did not carry the gene, could still serve as controls. Participants would undergo cognitive testing and neurological exams, donate blood and spinal fluid for testing, and have their brains scanned to measure physical structure as well as any amyloid deposits. After they died, DIAN scientists would study their brains at autopsy.

Morris wanted to know: What were the first abnormalities to appear in the brain of an Alzheimer's patient? How long before symptoms of dementia appeared did that happen? What biomarkers changed when, and in what sequence? By answering those questions, which mirrored the ones Reiman and Tariot were asking, Morris hoped to better pinpoint the optimal time to intervene with drugs.

He also wanted to compare the pathologies between mutation-triggered and run-of-the-mill Alzheimer's disease, to see how comparable the rare form was to the one that was stalking the general population. If they were able to find a drug that helped the mutation carriers, that comparison would theoretically help predict whether that success might cross over for everyone else.

"We know the mutation carriers are going to develop symptoms," he said. "Everyone sees the power of intervening."

Because of its size, Morris's study would set the stage for clinical drug trials, and he already had a team in place. The trial segment of DIAN would be led by Randy Bateman, a distinguished professor of neurology at Washington University's School of Medicine, as its principal investigator.

DIAN would grow to span fourteen research centers: three in Australia, one in London, two in Germany, and eight in the United States. Starting in 2012, one of those sites would be the University of Pittsburgh, and Bill Klunk's Pittsburgh Compound B would be used for the amyloid brain scans at all the sites. Through their relationship with Klunk, the DeMoes were also able to enroll in DIAN, thus expanding their research contributions internationally. Some DeMoes would be too far advanced in their disease state to qualify for the drug trials; but they encouraged their children to enter.

With her firstborn gone and her second son now living in a nursing home, Gail was growing increasingly more fragile, though she still had good days. She continued her lifelong habit of jotting down poems and ditties, and once she learned how to navigate her home computer, she became adept at social networking. A regular haunt of hers was Facebook, where even her grandchildren's friends wrote to her. Off-line, her own friends continued to drop by the house, just to check in and socialize.

On bad days, without the responsibility of caring for Doug, she did not get dressed or get out of bed. If she was awake and lucid enough to speak when Karla called to check on her, she talked about Brian.

"It's hard on that old lady," said Brian's son, Yancey, and he was right.

But the stress was also getting to Karla, who did what she could to help her mother from across the state. She set Gail up with a life-call service in case she fell and needed assistance while Dean was at work. She arranged for a housekeeper. She worked tirelessly, and even when Matt and her children told her to slow down, she could not. One of their rooms was filled with documents related to Alzheimer's to help her keep track of her family's journey. And though she usually managed to talk about her siblings matter-of-factly, she sometimes became overwhelmed at the thought that one day they would all leave her, and she would be the lone survivor of what had been a large and thriving family. She was not sure she would be able to function when that happened.

The most reliable help for Gail came in the form of Chelsey Determan, who adored Jamie's mother. In fact, the two women—young mother and great-grandmother—became the closest of friends. Gail was enchanted with Savannah, her youngest grandchild, and often referred to her as "the apple of Grandma's eye." She kept tea sets for her in Karla's old bedroom, so they could have parties; Gail kept Savannah for sleepovers, and the little girl grew into a precocious little spitfire with a butterscotch-colored ponytail, like her father in appearance but far more confident. When she visited her uncle Dougie in the nursing home, she played easily with all the elderly residents; she had none of the shyness that children sometimes display around old folks.

Chelsey looked after Gail as though she were her own mother, fussing over her prescriptions, pitching in whenever she was needed. She made the time.

Gail called her a godsend.

Jamie worked long hours and remained characteristically worried about life, but he doted on his little girl, calling her by her nickname, Nanna. He tried to be strict with her but was often disarmed by her sass. When he visited the University of Pittsburgh to contribute to the DIAN study, he and Chelsey always brought Savannah along with them, and she was a favorite among the staff there.

When a psychiatrist interviewed Jamie about his life, asking him what he did to keep busy, he didn't know how to answer.

"You mean besides work?" he said. The truth was, not much. He was assigned to a rig that operated around the clock. Depending on where it was located, he sometimes put in a fourteen- to fifteen-hour workday, and his schedule placed him on the rig two weeks in a row, then off two weeks. He was earning a six-figure salary, and it was helpful that he never had to schedule vacations, but it was brutal, dirty, exhausting work—not to mention dangerous.

But even during the spans when he had two straight weeks off, Jamie struggled to occupy his time. He blamed the oil boom for ruining Tioga's small-town atmosphere.

He didn't really talk about it much, but Chelsey could tell he was depressed. She also thought she detected some subtle lapses in his memory. You had to be around him daily to really sense it; he didn't get lost, didn't forget major events, nothing like what Brian and Doug were experiencing when their employer flagged them.

Depression is extremely common in Alzheimer's patients, particularly in the early stages of the disease. The National Institute of Mental Health established formal guidelines for diagnosing depression in someone with Alzheimer's, reducing emphasis on verbal expression but including irritability and social isolation.

Chelsey and Jamie had known each other for six years, but "if you ask Jamie, he always says three. It's fine," Chelsey said, laughing her rich, deep laugh. By now, an engagement ring winked on her finger, but they hadn't set a date. Nobody was really sure when, if ever, they'd actually marry. "Always. It's always 'three.' "

But sometimes, he had trouble recalling other details, too. Like their first night in Pittsburgh on that trip, when they'd taken Savannah to a Pirates game; they'd left before the end and watched the final inning on the hotel TV, but Jamie hadn't recalled that the next day. Or the trip they'd taken to the Minot State Fair: They had driven in and met up with a friend, but Jamie thought for some reason that they'd all driven there together.

He'd never been great about following story lines in movies or television shows, and he was tired all the time. He was never overly happy. Dean had told him he'd still be the same person, and he wanted so much to believe that. He longed to return to the man he had been before the diagnosis: the one who'd had the guts to persuade Chelsey to dance, who broke hearts across town and could party with his nephews Yancey and Tyler just as easily as he could with his older brothers. He just didn't know how to get back to that place in his life.

The research specialist asked Chelsey if circumstances were growing more difficult. Could she rate her own stress levels on a scale of one to ten?

"You can have a lot of stress, just dealing with it," said Chelsey, and her voice caught as she struggled to hold back tears. "I talk to Jamie's mom a lot."

She wondered if antidepressants would help Jamie, but she knew he would resist taking them. Like Dean, he eschewed medication. He didn't even drink much.

While antidepressants are one treatment for depression in Alzheimer's patients, researchers have found that other approaches—support groups, reengagement in favorite activities, celebrating small successes, exercise—also make an impact. Jamie said he would try. Reassuring the patient that he would not be abandoned was also important, and Chelsey made it clear that she would stand by him.

For the first time since his arrival at the research center three days before, Jamie finally relaxed when he settled into the big recliner they kept in the lab where they'd be scanning his brain. His arms were covered in bruises from all the needle sticks; one of the nurses apologized for how much trouble she had putting an IV line in his arm.

"Don't tell Chelsey or Savannah," he said, referring to the chair, because he wanted it all to himself. Then he remembered: They couldn't come in here, anyway.

He was in the Positron Emission Tomography (PET) Research Center, part of the Department of Radiology at the University of Pittsburgh School of Medicine. Because she was only five years old, Savannah was banned from the center; the research team was concerned about her exposure to radiation. In fact, Jamie wasn't allowed to hug her or pick her up after his brain was scanned with PiB and a radioactive form of glucose; he might still be shedding radiation. As a precaution, the research team had booked a second hotel room for Savannah and Chelsey to share, and Jamie would sleep alone.

PiB clung to the amyloid accumulating in Jamie's brain; the research team compared those physiological findings with the results of his written cognitive tests, answers to questions, and physical examination. They also measured how each part of his brain metabolized the radioactive glucose.

A staff member wheeled a canister down the hallway from the room where they'd manufactured a dose of PiB in Chet Mathis's beloved cyclotron. When the cyclotron was not in use, they covered it with a protective tarp; a cafeteria was located directly overhead, and syrup from the soda fountain had once dripped through the ceiling onto the cyclotron, giving Mathis fits.

A technician, his hands coated in gloves, removed the freshly made

PiB from its canister and injected it into the IV line that had finally been placed in Jamie's battered forearm. Jamie squeezed his eyes shut and grimaced. He hated this part, hated feeling abandoned in the scanner. He asked if someone could please check on him every fifteen or twenty minutes, so he at least had some point of reference to know how much longer he'd be there.

He'd made the same request the previous year, and nobody remembered to do it.

"Sometimes it feels like you're in there for hours," he said.

For most of the DeMoes, the research visit was an act of pure altruism. They knew they would not benefit from what they were doing—they were already past the point where experimental drugs might help them. That was part of what made their commitment so admirable: They didn't have much time left, perhaps only a handful of years at best—and they were still giving so much of it up to advance the science beyond their life spans. In Dean's case, he was taking time off from work that he needed to earn money for his family after he was gone. As symptoms advance, Alzheimer's patients struggle more with travel. For Lori, each visit significantly and permanently subtracted from her ability to reorient herself once she returned home, and her symptoms worsened with each trip. Nothing deterred them.

But for Jamie, the motivation was much more personal. In the few minutes before he had to climb into the hated scanner, he was thoughtful, almost philosophical.

"You want to know why I do it?" he asked. "Probably a lot for Savannah, or maybe in my mind, that they'll find a cure for me. It's different for me, being ten years younger than Dean [*sic*]. I'm hoping maybe something—even not, maybe, a total cure, but something, maybe, where I could get five more years out of my life."

His best friend, Rikki Rice, knew the main reason Jamie wanted to buy time.

"All he ever says about Savannah is that he hopes to see her graduate. If there's one thing he talks about the most about the whole family disease, it's seeing his daughter graduate from high school," Rikki said as he sat at the

Rig II bar in Tioga after a long workday. “His biggest fear is that he won’t be around, or not mentally stable.”

Suddenly, the thought of what lay ahead overcame Rikki, and the big man sat in a bar filled with roughnecks, his drink on the table in front of him, and began to quietly weep.

“I don’t want to see him like that,” Rikki said of his friend. “I don’t know how to prepare for it.”

Twenty-One

LANDSLIDE

ON OCTOBER 28, 2012, the DeMoes' cousin Dawn died at a Wisconsin nursing home. She was fifty. Though she had never been able to participate in the DIAN study, her brain was sent to the University of Pittsburgh, joining Brian's, so she could posthumously contribute to a cure.

Her daughters, Leah and Alayna, were part of the study; they wanted to do their part to combat the disease that had taken so much from their mother's generation. But they refused to allow Alzheimer's to change their life's course. Leah had a second baby in 2015. Later that year, Alayna would also start a family. The way Leah saw it, some families have a predisposition to cancer, or heart disease, or diabetes. It wasn't a reason to give up on life. She had fought too hard for this one. She vowed that she would be, to her children, the mother she never had.

Her father kept reminding her to keep her mind on the present, not the future.

"What's wrong with twenty-five great years?" he asked. "Be grateful for the time you have, and then move on from there."

• • •

Dawn's sisters, Robin Harvey and Colleen Miller, were also part of the DIAN research cohort.

Robin just adored the staff at the Alzheimer's Disease Research Center, most especially Bill Klunk, who always seemed to be looking out for her best interest.

Robin had known since 2006 that she had inherited the Alzheimer's mutation. She was saddened by Dawn's death, but she had little contact with Leah or Alayna and nothing but contempt for her sister Colleen. None of them ever outgrew their sibling rivalry; if anything, it deepened as their troubles, compounded by the disease, continued to mount.

Robin was married to her first husband long enough to have a son and daughter before they split up when the children were toddlers. A few years after her divorce, she met her second husband, Mike Harvey, with whom she had three daughters. But by 2004, that marriage also was dissolving.

On a late July weekend in 2004, Mike took the three little girls—then eight, six, and four—camping on the Eau Claire River in Douglas County. It was a peaceful, lazy summer day until a piece of the riverbank crumbled under two of the girls, and they plummeted into the water. Frantic, their father dove in to save them.

A group of high-school boys who had been idling on a nearby bridge pulled the girls to safety, but Mike and another bystander, who had also attempted to help, both drowned.

Two years later, Robin learned she had Alzheimer's disease.

"I was hoping—usually, one in the family doesn't get it," she said, perhaps thinking of her cousin Karla. "I mean, I knew my mom's mom had it, but I didn't really think about it, you know? I was just living my life. . . . Yeah, I cried."

She was furious with Leah for refusing to get a genetic test before having a baby, and she wasn't afraid to say so. She urged her own children to find out their mutation status before making that decision.

"There's other ways to have children," Robin said. "How many kids out there need a parent?"

Where her sister Dawn cried chronically, Robin's disease presented a different, more haunting symptom: She would laugh to fill empty spaces in conversation, almost indiscriminately: *Ha ha ha ha ha ha ha*, like one of the kuru victims that Dmitry Goldgaber's old boss, Nobel Laureate Carleton Gajdusek, once studied. When she talked about her worries over one of her younger daughters, who had largely stopped speaking since Mike's death, she'd laugh uproariously: *Ha ha ha ha ha ha ha*. When a psychiatrist at the Alzheimer's Disease Research Center asked her what her days were like when her kids weren't home, she giggled.

The psychiatrist asked Robin: Did she ever contemplate suicide?

"Oh, no," she insisted firmly. "I'm trying to live as long as I can."

Her two children from her first marriage were now grown; of the three she'd had with Mike, one was nearly finished with high school, and she believed Mike's brothers would take care of the other two when she no longer could. But it was painfully apparent that she had nobody like Karla to keep an eye on her. She had become so easily disoriented that the Pittsburgh staff monitored her with an electronic tag so she didn't accidentally leave the building and get lost; yet when she returned to Wisconsin, she would resume driving.

"It is what it is; what are you gonna do? Make the best of it while you can. It's a sad situation. And I want to make the kids feel safe and [I'll] try to be here. I'm sure I'll still be OK until they graduate," she said, sitting in a conference room at the Alzheimer's center and adding her chilling punctuation: *Ha ha ha ha ha ha ha.*

"Things happen for a reason, or people come into your life for a reason. And I feel so lucky that I get to be here. Because coming here makes me feel like someone's watching over me."

Dawn and Robin's youngest sister, Colleen, had always been the black sheep of the family. Her personality was an odd mashup of sharp contrasts: intelligent, but flighty; gregarious, but irritable; ambitious, but chronically rebellious. Growing up, she liked to compare herself to her cousin Dean, because they were both rule breakers; but unlike Dean, Colleen had never outgrown that trait, never settled down.

In her prime, she had owned a successful beauty shop, two cars, a snowmobile, and a house with an impressively large yard, all the product of her innate gumption and creativity. But by the time she was in her midforties, it was all gone; the business was closed, her equipment was in storage, and she was well on her way to becoming the crazy cat lady of Brule, Wisconsin.

"She's got the kindest heart, and you can wound her deeply with anything cruel," said Becky Vork, who grew up next door to the three Miller sisters and was Colleen's closest friend. "She has a really kind, kind side, but she comes off abrasive to other people if you don't really know her."

Becky would prove to be the rarest of friends, one who was truly there, even in the depths of the disease. Many of the DeMoes experienced the isolation of friends who disappeared when they developed symptoms, from Gail and Moe to Steve and Lori McIntyre. People promised to help, offered support, and then scattered. It was just too hard, seeing a middle-aged person in diapers, stuttering over the simplest words, exploding in anger over invisible slights.

But Becky was different. She was sterling, and she would do anything for Colleen, whom she considered a sister. When business began to taper off at Colleen's beauty salon, it was Becky who confronted her. Customers she'd had for more than a decade complained that Colleen was becoming inconsistent, irresponsible. She fell behind in her rent; she fell behind in her taxes.

"Be honest," Becky said to her friend. "What the hell's going on? Where's the girl that was kind of like my hero, running her own business, having her own house?"

Colleen was evasive. She blamed the drop-off in her business on the slumping economy that followed the attacks of September 11, 2001.

"Come on!" Becky said. "The world stopped getting haircuts? That's ridiculous. And I get that the economy plays into your business, but mostly it was people being driven away by her behavior."

After a one-night stand with an old boyfriend over Labor Day weekend in 2004, Colleen became pregnant. Though it was unexpected, Colleen welcomed motherhood, giving birth to a daughter in 2005.

Three years later, she was diagnosed with the mutation.

As Colleen's memory declined, she began taking first Aricept, then Exelon, a skin patch that was another form of cholinesterase inhibitor, preventing the body from breaking down the neurotransmitter acetylcholine—a temporary and not always reliable fix for Alzheimer's symptoms. But she kept that information to herself for a while; she didn't want people to think badly of her. A social worker friend helped her figure out how to pay her bills and negotiate some of the complexities of daily living; at home, she also relied on her little daughter to help her with small tasks and remember where she had misplaced things.

When she visited Pittsburgh for the DIAN research, Becky came with her, using vacation days she had carefully hoarded from her job. She packed extra clothes for Colleen, knowing her friend was too disoriented by Alzheimer's to bring what she needed. She monitored Colleen's bank account, replenishing it in $100 increments, checking her debit card for accidental purchases. She answered questions for the research staff about Colleen's history and current state of mind.

Colleen wanted Becky to assume guardianship of her daughter when the time came. Becky, who was entering her fifties and whose own son was grown, believed the little girl would be better off with someone younger, a biological relative. The child's father was not an option; he was unemployed and had been known to struggle with drugs. But a paternal aunt with children of a similar age was eager to step up. Becky helped design the plan to make that transition and checked in with Colleen's doctor and social worker to monitor the situation.

Like Karla, Becky sometimes had to be the bad guy with Colleen, the person who told her things she didn't want to hear—like the fact that her daughter would be better off with an aunt, whom Colleen disliked but Becky trusted.

"I will do anything I need to do for that kid and for Colleen," Becky said. "All I say to my husband is: I'm just trying to do what's right, every day."

Sometimes, she thought about their childhood: inseparable girls who lay on the Millers' kitchen floor, eating cookie dough and calling the local radio station, begging the deejay to play their favorite song: Fleetwood Mac's "Landslide."

She couldn't have imagined, then, how hard things would get: First Pat, then all three of Pat's girls. Better not to think about it, she thought. Better instead to focus on what Colleen was giving back to science than on what the disease was taking away.

"If I think about it too much, I'll die," she said.

Twenty-Two

SOMETHING TO SHOOT FOR

DEVELOPING A DRUG, any drug, for eventual use in human beings is a mind-bogglingly expensive proposition fraught with risk for the company involved. Some sources peg the median cost of creating a new pharmaceutical at roughly $350 million, but that figure rises into the billions if the company has brought multiple medicines to market in a decade, because for every success, there are many more failures, and the business has to carry the weight of each failure on its books. Part of the risk inherent to investment in drugs is that so many of them fail.

Of course, the payoff for success is equally large. When Alzheimer's researchers begin to lose faith, they look for inspiration to one of the most famous stories in recent pharmaceutical research: the discovery of Lipitor, the trade name for Pfizer's wildly successful statin.

In the early 1980s, drugmaker Parke-Davis spent eight years developing Lipitor to reduce cholesterol levels and prevent heart disease. At the time, nobody was sure if the right target was LDL (now called "bad" cholesterol), HDL ("good" cholesterol), or triglycerides (fats associated with buildup in artery walls)—just as nobody in the Alzheimer's field is sure if amyloid or tau is the right target, or if it's both.

At first, Lipitor didn't generate a lot of enthusiasm at Parke-Davis. Then known as CI-981, it functioned about as well as its competitors when used in animal models, but one rival drug was already on the market, and three others were already in large-scale human studies.

Given those circumstances, the company was reluctant to pay for expensive human trials but finally agreed to move forward because the patent on one of its top-selling drugs was about to expire, and the market for statins was expected to be so big that even a small share would make money—just as the market for Alzheimer's drugs is expected to be huge. With little else in the pipeline, Parke-Davis threw a Hail Mary on CI-981.

Twenty-four employee volunteers served as the drug's guinea pigs, and their results would make medical history: Despite the drug's lackluster performance in animal models, it was extremely potent in humans. Eighty milligrams of CI-981, the recommended dose for patients with very high cholesterol, beat any other statin by a 40 percent margin.

Even with such impressive results, Parke-Davis's wonder drug wasn't a lock for success. To convince the FDA to fast-track its approval for CI-981, now renamed Lipitor, its developers would have to demonstrate that the drug filled an unmet medical need. And since competitors were already available on the market, that argument was a tough sell.

The company found its answer in a group of South African children with a rare genetic disorder that hindered the body's ability to clear cholesterol. The disease left signature lumps of waxy deposits under the skin on their fingers. With levels topping three times that of a normal adult's cholesterol, the children were dying from heart attacks at an average age of fourteen. Other statins had failed to budge the condition.

But Lipitor did. Though it didn't cure all the stricken children, it packed enough of an impact to persuade the FDA to prioritize the drug. And it worked on the rest of the population, too, not just those who were genetically stricken. By 1997, Lipitor had been cleared for sale. For Pfizer, which acquired Parke-Davis, it was a true blockbuster, with annual sales topping $11 billion before its patent expired in 2011, allowing a cheaper generic version to enter the market. Lipitor became the best-selling drug in pharmaceutical history.

Now pharmaceutical companies large and small were looking for the next Lipitor. And it seemed as though Alzheimer's might be the right disease for just such a discovery.

All the pieces seemed to be in place: PiB allowed scientists a new window on the brain with images that would illustrate how the disease changed it, at least in terms of amyloid accumulation. No effective treatments existed, so when research suggested solutions that seemed like long shots, the government—driven by the pending health care crisis of aging baby boomers—was willing to listen. And in Alzheimer's, as with heart disease, pharmaceutical companies recognized a market of vast potential, one that might even surpass Lipitor.

The research field, as well as pharmaceutical companies, focused heavily on amyloid as a target. But as additional studies began to back up the tau theory, suggesting it was the primary driver, drug development was following suit, as was the push to image tau proteins the way PiB imaged amyloid. Some scientists, including Eric Reiman, believed that a successful Alzheimer's treatment might well be a combination of drugs targeting both amyloid and tau, just as combination therapies helped HIV patients.

Drugs were the key protagonists in human clinical trials; if they succeeded in showing a clinical benefit in Alzheimer's patients, researchers could apply for FDA approval. Several were in various stages of development, despite discouraging statistics. From 1998 to 2011, a whopping 101 Alzheimer's treatments had failed to reach patients. During that same time frame, three medicines won approval to treat symptoms of the disease, but their benefits were only temporary. With a record of three lukewarm wins and 101 losses, the smart investment play would have been to look for other, more promising diseases to treat.

Still, the industry soldiered on: Another ninety-three new medicines for Alzheimer's and other dementias remained in development. A drug that completely prevented the disease was the holy grail, but one that delayed onset by even five years would be considered a major victory, not only because it would buy valuable time for the patient, but also because it would signal that science was on the right track in targeting the disease's elusive underlying pathology.

There were signs that the history of repeated failures was taking its toll. A 2015 report compiled by the Dementia Forum of the World Innovation Summit for Health (WISH) warned that donors and big pharma had developed a case of "funding fatigue." Between 2009 and 2014, major drug companies cut the number of research programs into central nervous system disorders by half. As Eric Reiman warned, the field was not going to allow many more shots on goal; they had to get these drug trials right, or as right as they possibly could, because others might not follow. Drug companies were businesses, and they would invest in treatments that paid off.

The committee charged with selecting drugs for the DIAN clinical trials had to choose from fifteen different candidates nominated by drug companies. Committee members evaluated each based on the drug's safety, the mechanism by which it worked, its potency, and its stage of development. From 2010 through 2012, they categorized and prioritized the drugs until they whittled their original list down to two:

Eli Lilly and Company's solanezumab, nicknamed "sola," an antibody which bound itself to soluble forms of amyloid carried in the bloodstream before the body deposited it into plaques, and Roche's gantenerumab, which bound to amyloid in the clumping stage.

Solanezumab had already been tested in people with full-blown Alzheimer's disease, with disappointing results. But data analysis from those trials suggested that it slowed cognitive decline in patients with the mildest forms of the disease, indicating its potential as a preventative drug.

The trial's success would be measured incrementally. Of course, the ultimate goal was to stop Alzheimer's for all time, but that was a high bar; even if investigators found a wonder drug, they estimated it would take twenty years to show that presymptomatic patients did not develop the disease after taking the compound; it would take forty years to eradicate the disease. At a minimum, the trial hoped to show that the drugs conferred at least some subtle cognitive benefit that budged biomarkers.

"That's possible. That's something we can shoot for," said Randy Bateman, the DIAN trial's principal investigator.

To participate, volunteers had to be at least eighteen years old and anywhere from fifteen years younger to ten years older than their parent was when they first started showing signs of Alzheimer's disease. In cases like Dawn's daughters, Alayna and Leah, determining their mother's age of onset was difficult, since Dawn's strange behavior began in her thirties—a possible side effect of the antidepressants she was taking—but she was not formally diagnosed with Alzheimer's until years later. In cases such as theirs, doctors asked questions during the screening process to tease out a probable age of the disease onset in the parent. But since women participating in the study had to agree not to have children while they were taking the drugs, both Leah and Alayna opted out. They wanted children. Leah was already married with one child at the time she first participated in research.

The study's volunteer range also included people who were classified with mild cognitive impairment but were still healthy enough to potentially reap some benefits. Worried that Dean was close to missing his window of opportunity, Karla urged him to join before his symptoms progressed any further. Already, it was too late for Lori and Doug. Though Dean hoped to complete his drug tests in Pittsburgh, where he was comfortable with the staff and the setting was familiar, he agreed to go to St. Louis instead, because Washington University was set up to accept trial participants several months before Pittsburgh. The trial enrolled a total of two hundred people.

Participants brought either a family member or a friend to their visits to provide an outsider's perspective about the real-world impact the drugs were having on their daily lives. The study also included a placebo group, meaning some participants wouldn't get a real drug at all; a computer randomly assigned the placebo to volunteers, although people who did not carry the mutation would not get the drugs, whether or not they knew their status. Because DIAN was testing two drugs at the same time, participants who were mutation carriers had a 75 percent chance of getting an active drug instead of a placebo.

After the initial screening visit at one of the DIAN sites, once a month a team of visiting nurses would administer the drugs to volunteers in their homes, either by injection or IV line. Every three months, participants underwent an MRI scan as a safety screen, designed to detect any brain swelling, a possible side effect; every six months, they completed cognitive testing on computer tablets. And once a year for three years, they went back to a DIAN site—such as Washington University or the University of Pittsburgh—for a physical, blood samples, brain scans, a spinal tap, additional cognitive testing, and more drug dosing.

Even in the early days of DIAN, Bateman and John Morris coordinated efforts with Eric Reiman and Pierre Tariot's Banner Institute, which was working to launch a drug trial of its own in the Colombian population championed by Francisco Lopera and his American partner, Ken Kosik. Their cooperation wasn't just the product of professional courtesy; it was a necessity. Competing against one another would only cause failure in all the trials, Bateman predicted. There were simply too few mutation carriers to go around, and each failure would dampen an already skittish market. Pharmaceutical companies were not going to invest indefinitely.

"We're only really going to get this time to do this," he said. "There's a place for competition, but there's a time when competition hurts."

Moreover, wasted time would most greatly impact the volunteers, who could not afford to age beyond the point where the drugs would be able to help them.

"There's no chance that they would escape the disease if they have the gene," Bateman said. "It's quite a desperate situation for them."

In Oklahoma, forty-two-year-old Sherry DeMoe Pickard, Jerry and Sharon's oldest daughter, prepared to travel to St. Louis for the clinical trials associated with the DIAN study. She decided against genetic testing, but her family supported her decision to move forward with the drug trials. Her six daughters were all still at home—the oldest twenty, the youngest seven.

The testing was moot anyway; Sherry's memory was starting to fail. The woman who had once planned to become a teacher could easily recall some details—such as the new curriculum she was using to homeschool her children, and the color of the Volkswagen bug her father had refurbished for her sixteenth birthday—but had trouble with others, such as how many years she had been married, how many sisters her husband had, or how old she was now.

When she went to Washington University for her baseline visit in 2013, she was pleased to be able to contribute to Alzheimer's research, even though the testing, poking, and prodding exhausted her. Less than a year later, she could not remember how long she had been there, or whether the doctors had taken her blood or conducted a spinal tap. Her family encouraged her not to worry too much about it, and she did her best to comply.

But while she did not dwell too much on the disease, or the uncertainty of her own fate, she did believe it was important to contribute to knowledge that might benefit the next generation.

"I have a lot of daughters," she said. "By doing all of this, it'll help for their future, doing what I'm doing."

Her mother, Sharon, accompanied her to St. Louis to serve as her study partner.

"There's times when I think, 'I can't do this again,'" said Sharon. "But I know I can. And I will.

"God has purposes for everything. We don't always know what and why. And sometimes we find out, and sometimes we don't."

She thought back to when Jerry was sick—a time when she was so distraught that she contemplated suicide. She lived through it, and now she knew why.

"I think the purpose is for me to be here for them, for her. I think that was the main purpose. We got through all the other, and I can't say this is going to be bigger; I don't know . . . I don't know how this is going to be with a child."

Part FOUR

Few will have the greatness to bend history itself, but each of us can work to change a small portion of events, and in the total of all those acts will be written the history of this generation. It is from numberless diverse acts of courage and belief that human history is shaped.

—Robert Kennedy, June 6, 1966

Twenty-Three

THE SILVER TSUNAMI

AS BATEMAN'S TEAM prepared to begin its drug trials, other scientists were simultaneously working on the bigger picture: how to prevent Alzheimer's in the general public. By marrying the science of the two populations together—those genetically stricken and those who spontaneously developed Alzheimer's without a mutation—the doctors hoped to solve what was shaping up to be a massive health care crisis.

Paul Aisen, the Baptist believer who headed the Alzheimer's Therapeutic Research Institute for the University of Southern California, was instrumental in that effort, along with his longtime collaborator, Harvard professor Reisa Sperling, who served as the study's principal investigator.

Their study attempted to answer this critical question: How would older people from the general population, not mutation carriers, respond to anti-amyloid drugs before their memories began to fail?

The two had met through the Alzheimer's Disease Cooperative Study, a consortium of eighty different clinical sites in the United States and Canada that formed in 1991 as part of a partnership with the National Institute on

Aging. Aisen, veteran of many ADCS efforts, led the cooperative from 2007 through 2015.

When scientists want to research Alzheimer's treatments, particularly those that don't interest the pharmaceutical industry, the ADCS helps drive those studies. For example, it successfully debunked the myth that ginkgo biloba, one of the top-selling herbal remedies in the United States, improved memory for Alzheimer's patients.

The DIAN study was testing whether the Lilly drug solanezumab, which flopped in people with more advanced Alzheimer's, worked better in people who either had no symptoms or were mildly impaired. DIAN would target people with all the known mutations: PS1, PS2, and APP.

Solanezumab carried enough safety data to convince Reisa Sperling that it was viable for use in the study she and Aisen were planning. Even if it wasn't a perfect solution, she thought it would likely answer some questions that she had been dying to ask, such as: What true risk of impairment does amyloid represent? After all, some people carried amyloid in their brains yet functioned normally, which researchers now knew from imaging amyloid in a living brain with PiB.

Studies in older people had the potential to relate more directly to the garden-variety Alzheimer's that had started to close its grip on the world's population. As baby boomers aged, society was becoming increasingly worried about the potential consequences if no means of prevention could be found.

Several reports about those concerns labeled baby boomers "the silver tsunami" because of the possible devastation their health demands could wreak on an already burdened system. In 2013, a study published in the *New England Journal of Medicine* predicted that the $109 billion cost of caring for Alzheimer's patients in 2010—not counting informal care costs, such as services provided by family members to take care of the person—would more than double by 2040. Sperling and Aisen had their work cut out for them.

Also working in the older population were Pierre Tariot and Eric Reiman of the Banner Institute, in an effort separate from their research into the Colombian *paisa*. They shared information with Aisen and Sperling, as well as the DIAN group and, for a time, Allen Roses, the tauist from Duke who first identified variants of the ApoE gene on chromosome 19 that heighten the risk of Alzheimer's disease in the general population. (Roses died in 2016.) The scientists in each study compared notes: What's working? How do we standardize what we're collecting so we can compare results? How are we measuring cognitive function? How do we disclose any of our findings to the public? The combined efforts created important momentum, with each study contributing a different perspective to the field's knowledge base, helping to move any future prevention studies forward.

The two trials focusing on older people were, like the mutation studies, driven by genetic predictors. The Banner Institute's trial sought to treat 650 people between the ages of sixty and seventy-five, most from the United States. Instead of an autosomal dominant mutation, they carried two copies of the ApoE4 gene variant identified by Roses. Though not as rare as the mutations that plagued the Colombians, the DeMoes, and the Noonans, the condition—which affects about 2 percent of the general population—was still potentially lethal. Researchers initially believed carrying two copies of ApoE4 represented a 90 percent chance of developing Alzheimer's; but subsequent risk estimates have lowered that number to 58 to 68 percent, or about the same risk that women with the BRCA1 gene mutation—including actress Angelina Jolie—have for breast cancer.

People who only carry one copy of ApoE4 have a lower risk of getting Alzheimer's—roughly 30 to 45 percent—but there are more of them. About 25 to 30 percent of the general population carries one copy of ApoE4. Reisa Sperling and Paul Aisen widened the parameters of their research in the general population to include these people as well, ranging in age from sixty-five to eighty-five years old. They named their study Anti-Amyloid Treatment in Asymptomatic Alzheimer's, or A4 for short.

The A4 study hoped to enroll a thousand people. Sperling and Aisen started by recruiting their subjects, whose cognitive function was completely normal at the beginning of the study, but in whose brains amyloid plaques were accumulating. At fifty-six different sites across the United States—plus four in Canada and one in Australia—about half would receive a placebo, and half would be dosed with solanezumab. An additional five hundred people with no amyloid plaques would serve as a second control group so Sperling and Aisen could compare their results to the normal aging process, a step that would help them better understand the role amyloid played.

It was a huge undertaking, one that defied FDA tradition. To approve a medication for use in Alzheimer's patients, the agency had always required drug trials to include two criteria for success. The first was an effect on some measure of cognitive performance, such as a memory test. But sometimes, test scores are deceptive; a patient can ace a test but still stumble over daily tasks that the results didn't capture.

So the FDA also required drug trials to measure "clinical meaningfulness," or a sense of improvement as judged by the patient, a caregiver, or a doctor. Aricept and other drugs that doctors routinely prescribed for Alzheimer's had won FDA approval by showing some benefit in both these measurements—cognition and clinical meaningfulness—even though the drugs didn't delay the disease, or even slow it down.

The problem was that these measurements seemed impossible to achieve when testing drugs on people in the "preclinical" stages. Someone who appears normal won't struggle with memory or daily tasks, so how can researchers measure an improvement?

But a surprising answer came out of the University of California, San Francisco. Its study analyzed about four hundred people diagnosed with mild cognitive impairment, two hundred with mild Alzheimer's, and two hundred who were considered completely normal.

What they found was startling: About a third of the "normal" group's members were actually spilling brain amyloid peptides into their spinal fluid—something that science would expect to see in an Alzheimer's patient, as did Trey Sunderland and Susan Molchan, but not a healthy person—and PET scans revealed small signs of brain degeneration. Aisen took those results as yet another sign that "amyloid is not just an innocent bystander."

But the study also suggested, as many had long believed, that the Alzheimer's brain was subtly changing years before anyone would notice a problem. Some brain scans did show degeneration—specifically in the hippocampus, a sea horse–shaped region nestled in the middle of the brain, responsible for forming long-term memories, similar to a computer's hard drive.

However, Aisen realized there was already a much simpler tool, one he'd previously dismissed as too primitive to be useful in patients whose symptoms were virtually invisible.

The Mini Mental State Examination, known to clinicians as the MMSE, is the most commonly used test for people experiencing memory problems. Scored on a thirty-point scale, it takes about ten minutes to complete and asks simple questions, such as the time and date, repeating back a list of words, or basic math calculations, such as counting backward from one hundred by sevens.

Since its introduction in 1975, little about the MMSE has changed. A score of 27 to 30 indicates that the person is normal, while the scores of Alzheimer's patients generally range from zero to 24. The first person ever dosed with PiB scored 25.

People with mild cognitive impairment, that gray area between normal and full-blown Alzheimer's, usually score between 24 and 30. But the MMSE is so imperfect that even people with dementia can sometimes score a perfect 30.

Julie Noonan Lawson thought the test, which she referred to as "their mini-mental thing," was silly. She recalled how her afflicted siblings—particularly Fran, the family brainiac—used to memorize the MMSE so they could outsmart it. One brother, asked to remember a grocery list, would categorize it and spit it right back out.

Yet the test was actually more useful than it seemed. Aisen realized that a series of MMSE scores separated by time changed consistently in people with brain amyloid who were otherwise normal. The changes were tiny, fractional, and still in the normal range. But they persisted.

Over a two-year period, someone with preclinical Alzheimer's would drop about a half point on the test, a difference too small to be important in one person; but when it happened in two hundred otherwise normal people, it was significant. Aisen and Sperling added this measurement to their arsenal.

The FDA also agreed to make its rules more flexible, thanks in large part to the efforts of Russell "Rusty" Katz, who worked for the FDA's Center for Drug Evaluation Research. It was Katz who had created the definition of "clinical meaningfulness" in the early 1990s. He realized that researchers needed more leeway if they wanted to test a drug in patients who were still apparently normal.

Before he retired in the summer of 2013, Katz wrote of his support for looser guidelines relaxing the clinical meaningfulness requirement. As soon as the proposed new policy appeared in the *New England Journal of Medicine*, it received support from pharmaceutical companies, whose funding was critical for such research. When the requirement was relaxed, drugs that treated people before they became symptomatic stood a much greater chance of winning FDA approval.

Although the FDA was willing to allow drugs to go to market without demonstrating that they improved a person's daily function, follow-up studies would have to show a benefit as time progressed.

Aisen believed this new model of collaboration among government regulators, academic researchers, and pharmaceutical companies was key to achieving any kind of breakthrough.

"It's too big a problem for anyone to tackle alone," he said. "What is striking is: Despite the fact that this field has utterly failed for over a decade, everyone is very optimistic."

Reisa Sperling wasn't expecting a home run from A4, but she was hoping for a glimmer of success. Maybe the trial would show a tipping point for amyloid—a level of accumulation after which people begin to lose their memories. If so, doctors might know when they needed to intervene.

And then there was the question of tau, the protein that made up tangles—not just in the brains of Alzheimer's patients, but also in people with sustained head trauma, such as football players and other athletes who played high-impact sports.

"I'm just amazed that we are still fighting about whether it's amyloid or tau, because it's so clearly *both*," Sperling said. Her husband, Keith Johnson—a radiologist and neurologist who also taught at Harvard Medical School—coined a phrase: "Amyloid pulls the trigger, but tau is the bullet."

To Sperling, the debate was better framed as a chicken-and-egg question: Which comes first, the amyloid or the tau? Does one beget the other, or are they both separate streams of the disease process?

Even though Aisen firmly believed that amyloid drove the disease, he pointed out that once it hits a certain point, its levels plateau, even as the brain becomes progressively more damaged. Tau, on the other hand, did increase quite a bit as the brain lost substance and function. He, like Sperling, wanted to know more.

To help answer those questions, they added yet another resource to their study: T807, a radiotracer that did for tau what Bill Klunk and Chet Mathis's Pittsburgh Compound B did for amyloid; it allowed doctors to see tau in the living brain through PET scans. Johnson, Sperling's husband, was an early advocate of T807. He traveled to conferences to speak about what he found in patients who were scanned using the compound.

For example, in November 2013, he told colleagues at the Clinical Trials in Alzheimer's Disease meeting in San Diego that in patients where T807 showed accumulation of tau in the brain, memory was impaired. But the patients could still retrieve words and perform the brain's so-called executive functions, such as keeping track of time, making plans, or evaluating ideas.

Increasingly, scientists began looking at the likelihood that salvation lay in treating both plaques and tangles. Sperling compared a brain overwhelmed by Alzheimer's to an overflowing bathtub: To fix the problem, you have to first turn off the spigot (amyloid), then empty the basin (tau).

Her hope was that her study, combined with the work of others, would help pinpoint which proteins were responsible for which symptoms. About half the participants would be scanned for the presence of abnormal tau under Johnson's leadership. Already, Sperling was thinking ahead to the possibility of combination therapies, one attacking amyloid, the other attacking tau.

"That's my holy grail," she said.

Perhaps it would be the combination that could deflect the silver tsunami, although Reisa Sperling found that particular metaphor wholly inadequate.

"A tsunami is a single wave. We're talking about wave, after wave, after wave of people developing risks for Alzheimer's," she argued.

Without meaningful intervention, it was a crisis powerful enough to overwhelm the country's health care system, to cripple the infrastructure of American society. It was a thought powerful enough to drive her relentlessly through the emotionally draining prospect of disappointment. From the moment she launched one clinical trial, she thought about the next step forward, and the next, and the next.

Even then, with every volley, she worried that she was going to fail. When Sperling was in medical school three decades earlier, her grandfather had developed Alzheimer's, inspiring her to find a cure before the disease could claim her father. She was too late; her father, a college chemistry professor, was diagnosed with Alzheimer's in 2014.

"Sometimes I feel like we're ahead of ourselves, and sometimes I feel like we're terribly behind," she said. "I'm worried that I'm losing the battle."

Jerry and Sharon DeMoe in an undated photo with their older daughter, Sherry. They also had a younger daughter, Sheryl. The family settled in Oklahoma, where Jerry taught diesel mechanics. SHARON DEMOE

Lori and her daughters (left to right): Jessica, Robin, and Chelsey. They traveled around the country following her husband's job with Union Pacific Railroad, even living in a bunk car for a while. JESSICA MCINTYRE

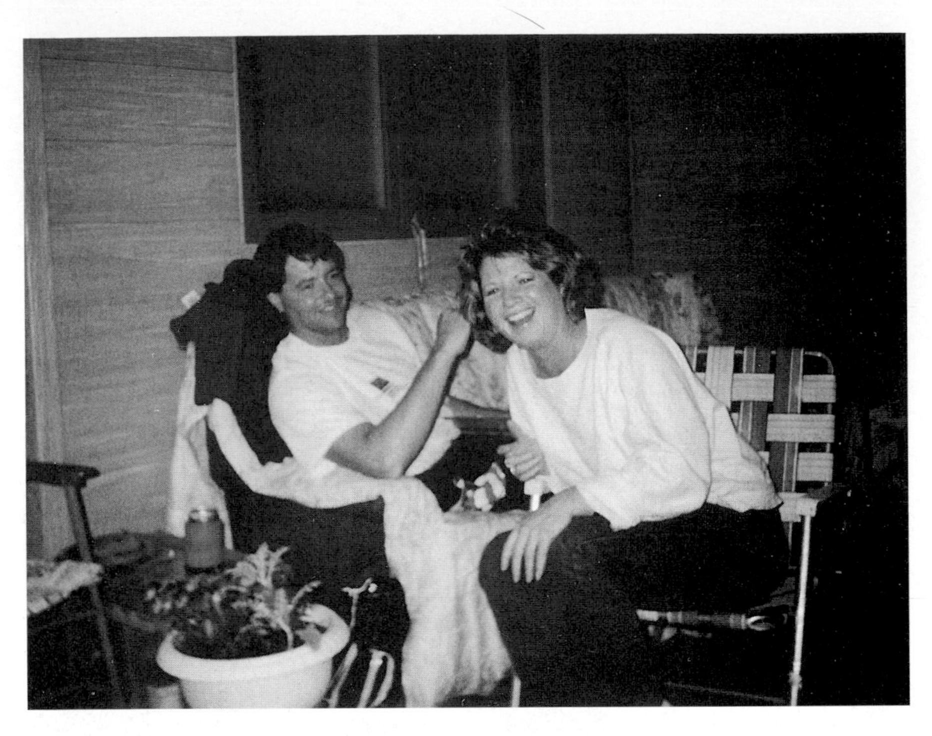

Dean DeMoe and his cousin Colleen in 1999. Both were successful in business and shared a rule-breaking spirit. DeMoe family

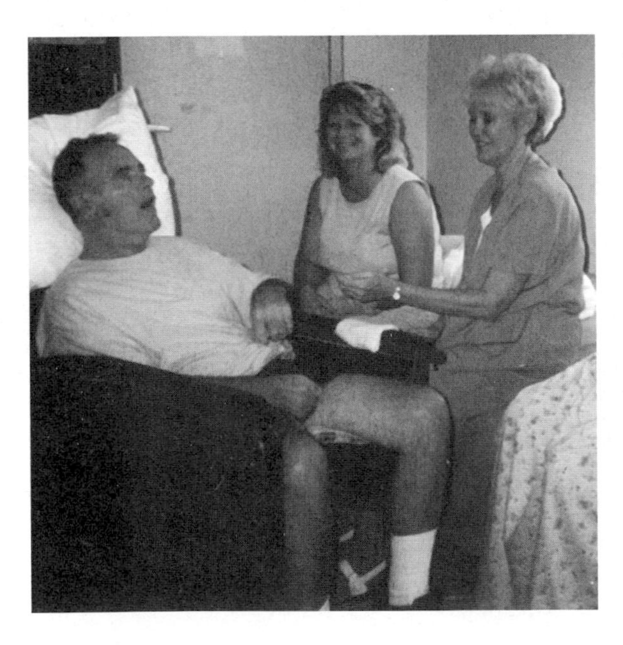

Jerry DeMoe in 2003 with Sharon still by his side. In the background is Colleen Miller, the youngest daughter of Jerry's sister, Pat. At the end stages of his life, Sharon brought him home to care for him and put on a brave face. But privately, she was desperate. She found comfort in talking to Jerry's sister-in-law, Gail, who knew what she was enduring. Sharon DeMoe

The McIntyre family, November 2007. From left: Chelsey, Jessica, Lori, Steve, and Robin. Though her family believed otherwise, Lori was convinced that she had the Alzheimer's mutation. Before her diagnosis, she met with her closest friends and made them promise to watch over her daughters when she no longer could. JESSICA MCINTYRE

Chet Mathis (left) and Bill Klunk from the University of Pittsburgh, whose Pittsburgh Compound B allowed science to view the progression of Alzheimer's in a living brain for the first time. Klunk would later play a critical role in linking the DeMoe family to research testing experimental prevention drugs, and family members consult him often for advice and counsel. RIC EVANS

In the 1990s, Kenneth Kosik (left), then a Harvard neuroscientist, teamed up with Francisco Lopera (right), a neurologist from Colombia who discovered the largest known extended family who carry a genetic mutation for Alzheimer's disease. Between them is Lucía Madrigal, a member of Lopera's research team. Together they worked tirelessly to convince the rest of the scientific community of the value this family represented for research. KENNETH KOSIK

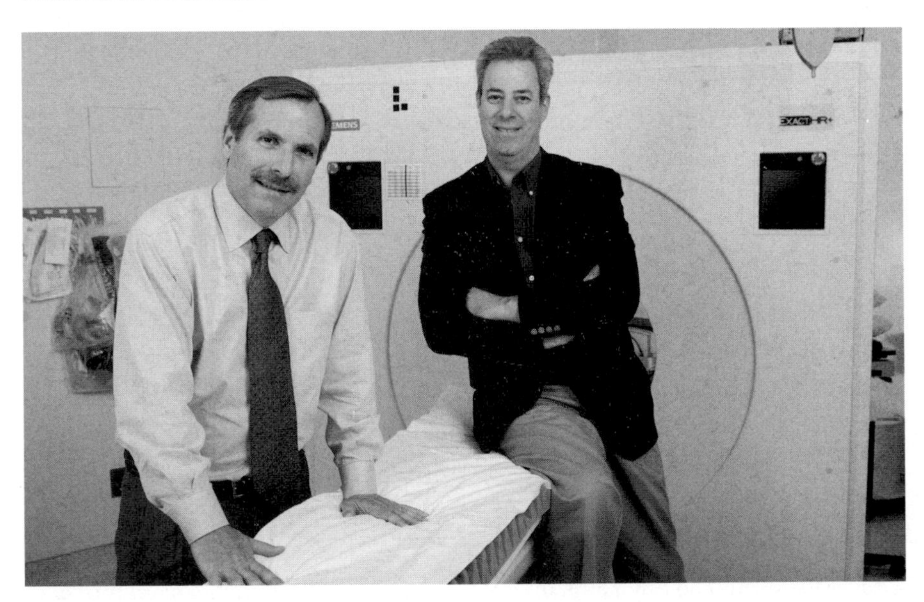

Pierre Tariot (left) and Eric Reiman of the Banner Alzheimer's Institute in Phoenix. Their ambitious plan to detect and attack Alzheimer's before a person develops symptoms led to a collaboration with Ken Kosik and Francisco Lopera to test a prevention drug in the Colombian families. BANNER ALZHEIMER'S INSTITUTE

Dr. Randy Bateman, head of the clinical trials unit of the Dominantly Inherited Alzheimer's Network (DIAN). This international effort is testing experimental Alzheimer's prevention drugs in the DeMoes and other families who carry the rare autosomal dominant mutations that cause the disease. WASHINGTON UNIVERSITY

Gail DeMoe surrounded by her grandchildren, circa 2005. Clockwise from top left: Jennifer (Doug's daughter); Lindsey (Dean's daughter); Yancey (Brian's son); Kassie (Brian's daughter); Jessica (Lori's daughter); Robin (Lori's daughter); Chelsey (Lori's daughter); Amber (Karla's daughter), holding McKenna (Dean's daughter), and Tyler (Dean's son). DEMOE FAMILY

Jamie DeMoe and
his daughter, Savannah.
DEMOE FAMILY

Brian and his daughter, Kassie, at her college graduation. DEMOE FAMILY

Gail (top) with her children. Clockwise from top right: Doug, Karla, Dean, Lori, Jamie, and Brian. Joni Syverson

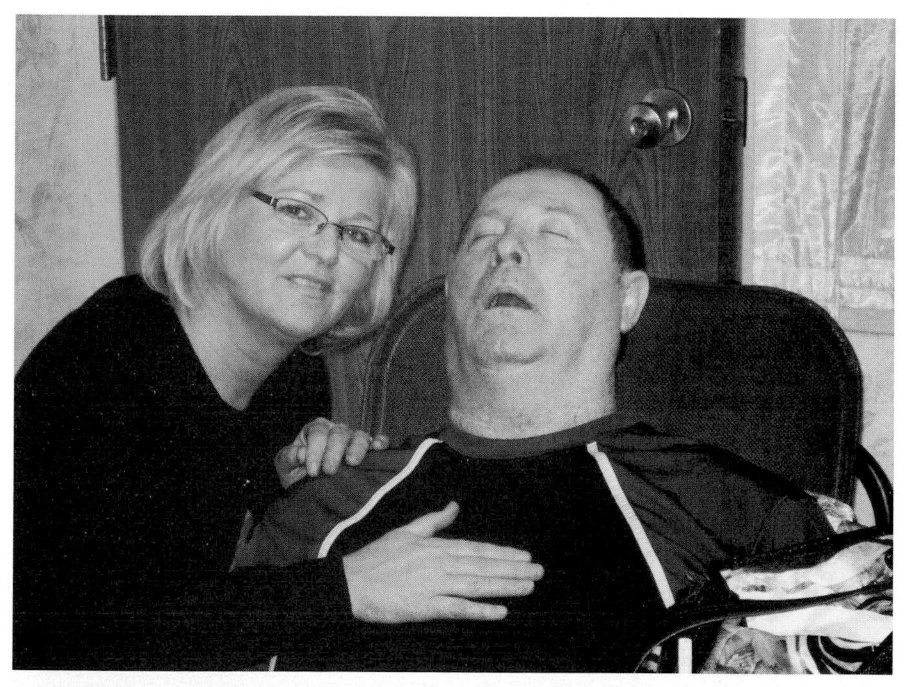

Karla and Brian during the end stages of Brian's fight with Alzheimer's disease. She dedicated her life to giving meaning to her family's generational ordeal. DeMoe family

Dean visiting his best friend, Monte Olson, in Australia, spring 2015. DEBRA DEMOE

The DeMoe family enjoyed a special connection with Bill Klunk, who often joined them for dinner when they came to the University of Pittsburgh to participate in Alzheimer's research. Left to right: Dean; Karla's daughter, Amber; Karla; Dean's wife, Deb; and Dr. Klunk. DEBRA DEMOE

Twenty-Four

THE LUCKY ONES

DOUG'S MOVE INTO the Tioga nursing home marked another grim milestone in Karla's life. Coupled with Dean's decision to return to the oil fields, and following closely on the heels of Brian's death, she was reminded more than ever that she would soon be the only sibling left, and she was cracking under the strain.

Compounding her stress levels was the fact that she was now working two jobs. By day, she worked as an administrative assistant at a college; by night, she moonlit at a service that typed legal paperwork. There were days when she worked from 8:00 a.m. to 10:00 p.m. Like Gail, she developed tremors in her hands, and when she was home, she cried; sometimes, she even had to leave work early to regroup.

It helped to talk to Gail, which she did constantly—not just about the disease, but about any little details that came up in the course of her daily life. And Gail unburdened herself, too, both of them recognizing that they shared an understanding of how hard it was to survive the death of people they both loved.

"We'd dump all our problems and then say, 'Well . . . have a nice day,'" Karla said.

Karla was turning fifty, but she had realized a fundamental truth: All your life, you will find a time when you simply need your mother.

In the summer of 2008, twenty-three-year-old Chelsey McIntyre, Steve and Lori's youngest, was facing the toughest dilemma of her life. She was a student at the University of Wyoming and had just discovered she was pregnant.

She had been dating a fellow student, Jeremy Francom, who had graduated and planned to move to an Indian reservation to become a teacher. The relationship was fizzling out when, about a week before Jeremy was scheduled to move, Chelsey found out she was expecting.

"It was very hard," she recalled. "It took a long time for me to come to terms with it."

Not knowing where else to turn, she called the one person who always seemed to know what to say: her grandma Gail. Predictably, Gail—lover of all children, especially babies—was thrilled at the news. Her enthusiasm bolstered Chelsey's courage, and finally, the prodigal daughter called Lori.

"She was upset at first, and then I think it finally set in: We're going to actually have a grandbaby. That's what she wanted," Chelsey said. Lori's wish was coming true; she would know at least one grandchild before Alzheimer's took away her ability to connect.

Chelsey's bigger concern, she confessed to her mother, was telling Steve. He was adamant that she finish college, and she just couldn't bring herself to tell him about this complication. So Lori broke the news. For weeks, there was nothing but silence from Idaho, where Steve and Lori were now living.

"I've had time to think about this," he said, when he finally called Chelsey. "I wanted to process it before I said something I didn't mean. I just really needed to think about it."

And then he told her: "I really think this happened for a reason."

Though Jeremy was committed to raising his daughter, Chelsey didn't think they should marry just because they had a child. They remained on friendly terms and stayed involved with each other's extended family, and they worked hard to coordinate their parenting styles, even when he moved three hours away to begin his teaching career.

Claire Kennedy Francom was born in March 2009. As she grew, she and her cowboy boots were inseparable. She divided her time between her mother's home and her father's, despite the distance. When she was with Jeremy, she loved to Skype with her mother; when she visited Steve and Lori, she bossed her grandmother around, as though they were the same age. Emotionally, because of the disease, they often were.

With Claire's arrival, Chelsey began to question whether she wanted to learn her own genetic status. Knowing she could have passed the mutation along to Claire, she decided to wait. Her own mother's diagnosis had been devastating, and she was already worried. Adding to the burden of young motherhood seemed like a bad idea, and both her parents were adamantly opposed to any of their daughters finding out their status, fearing the knowledge would burden them unnecessarily in their young adulthood.

"It's hard to make that decision," Chelsey said. "Do you have kids no matter what? You need to live your life as normal as possible. Or do you think about the next generation?"

The notion of parenthood was encroaching more each day on Chelsey's cousins. Following seven years of college, Dean's daughter Lindsey graduated with a doctoral degree in physical therapy. After some initial hesitation, she enrolled in the DIAN study at the University of Pittsburgh. She waited until she was sure that she did not have to learn her genetic status to participate.

"For right now, I'm perfectly content not knowing," she said. "If different treatment options become available, I might reconsider. But it's still at the point where finding out does not benefit me in any way, and it's not worth the emotional distress."

She became engaged to her boyfriend, Jon Sarkilahti, and together they bought a house in Minnesota. She knew she wanted to have a family, and she worried about how the mutation might affect those plans.

"Should I even have any kids? I struggled with that for a very long time," she said. "Knowing that the gene is in the family, would that be wrong of me, bringing a kid into this world, knowing that they had a fifty-fifty chance of having it? Or even if they don't have it, watching me struggle with it if I do have it?"

The question of children continued to dog the DeMoes: Did they have a moral obligation to discover their status? Hearing the scenario, bystanders often assumed the answer was obvious: Yes, anyone capable of childbearing, or of fathering children, should find out. If they tested positive, they should not have children.

But the day-to-day reality for someone who actually lives that situation is far less black-and-white. Deb DeMoe, who turned so often to her faith for answers, consulted her pastor. He pointed out that any number of circumstances or illnesses could take children from their parents: Accidents. Cancer. Meningitis. *Watch therefore: for ye know not what hour your Lord doth come*. The only difference in the DeMoes' case was the fact that they had some warning.

Though Lindsey decided against knowing, it was important to her that she contribute to a treatment, even if she never needed it herself.

"After Brian passed away, I wanted his life, and my grandpa's life, and all the other family members that had gone through the disease to be worth something, and to be able to leave a better legacy than how they passed away," she said.

A few months after Brian's death, Karla's daughter, Amber Hornstein, was married in an elegant church wedding in Fargo, attended by three of the family's littlest members—Savannah, Claire, and Kassie's daughter, Brianna—as flower girls. The entire extended family gathered to celebrate, including the cousins from Oklahoma and Wisconsin, and Gail was thrilled.

Slender, blonde, and blue-eyed, Amber made a stunning bride. She was working as a physician's assistant when she married, and a year later, she would enter medical school. She had been in high school when her uncles and aunt started getting diagnosed.

"It was the second or third [diagnosis] when I started to pick up that wow, this was serious," she recalled.

Like the rest of her family, Amber was worried about her mother.

"There are times when I think she would rather have the diagnosis," and spare one of her siblings, Amber said.

Her father, Matt, did not doubt that this was true.

"She's always thinking about it, always talking about it," Matt said. He summarized Karla's philosophy this way: "This is my fate. This is my job. This is my responsibility now, to take care of my family."

Karla was always at her happiest when she was surrounded by twenty relatives in a single room. "She's never going to lighten her load, until the day she dies," said her son, Cole. "She will never stop."

For a time, her daughter Amber had toyed with the idea of working for Bill Klunk, shadowing him.

"I live it every day," she said of the disease. "More than anything, I would love to find a cure."

Though she gave up the idea of working in Pittsburgh, she did put her medical training to use within the family, checking up on everyone's prescriptions and helping them interpret information. She believed her experiences would help to shape her into a better doctor.

One night at Grandma Gail's house in Tioga, Amber was sitting up talking with her cousin Chelsey McIntyre, who broke down crying and confessed something she hated to admit: She resented Amber and Cole for being free of the disease, for never having to watch their mother lose her mind, for never having to worry about passing the mutation on to their own children.

"I was shocked—so shocked," Amber said. "I know I'm lucky."

But thoughtfully, she added: "I guess I didn't realize *how* lucky I was until that moment."

And though some of her cousins had already been cleared, in the back of Amber's mind, a terrifying thought smoldered: "We're having all these no's, but I feel like a yes is coming on," she said. "It makes me physically ill to think about it."

After the champagne toasts were given at Amber's wedding and the dancing was done, after the plates were cleared away, the extended family gathered at Karla and Matt's house, including Sharon and both her daughters, Sherry and Sheryl, who had made the trip from Oklahoma. It was then that Karla realized Sharon's older daughter, Sherry, had the disease. Amber's premonition had been correct.

In between working, planning the wedding, and handling the aftermath of Brian's death, Karla had compiled a series of family photos into a slide show that she titled "Too Early, Too Many." It brought back the memories that had formed the mosaic of their lives: Moe holding Brian, as a baby, in his own mother's house; the DeMoe siblings, first as clean-cut, *Brady Bunch*–era youngsters, later as shaggy-haired teens; cousins and babies, aunts and uncles. Here were Dawn, Colleen, and Robin Miller as little girls in dresses Pat had sewn. There were Jerry and Sharon, standing together and smiling, and later, Sharon, smiling bravely as her husband lay in frail repose beside her.

The family sat, transfixed by the scenes from their life; many began weeping. The juxtaposition of the occasions, this one so close to the joy of the wedding, reminded them of the obligation that fate had assigned to them: Tragedy would intertwine with their blessings until they found the thread that would lead them out of the labyrinth.

Twenty-Five

FOLLOW THE SCIENCE

IN 2011, WHILE the DeMoe family regrouped at Amber's wedding, about fifty of the Colombian *paisa* were traveling in groups to the Banner Institute in Phoenix for brain scans that marked the beginning of what Lopera had long sought for his families: a sliver of hope. Eleven of the *paisa* already had symptoms; nineteen had no symptoms but carried the mutation that would guarantee the disease; and twenty were noncarriers. They were anywhere from twenty to fifty-six years old, and for many of them, the trip to Phoenix was the first time they had ever traveled outside of their immediate home turf or sat on an airplane.

Tariot and Reiman welcomed the group, most of whom they'd met previously on their trips to Colombia. They asked their guests: What would make this trip more comfortable for you?

The answer was unanimous. The Colombian kindred, to whom family was paramount, wanted to visit the doctors' homes. They wanted to see how these men lived, eat the food they ate, meet the people who were important in their lives. If they were truly going to trust these men to lead them out of this ancestral curse, they wanted to know whose hands they were grasping. And so it came to be that Pierre Tariot and his wife, Laura, welcomed groups of the *paisa* to their gracious, Southwestern-style home. They shared food and drinks, mingled, visited.

"We remembered what you told us, so welcome to our home. My house is your house," Tariot said. "You asked what's important to us, and you've asked why on earth we have your pictures on our walls at work."

The family members looked at him expectantly.

It was to remind him, every day, what his work was for, he told them. When he was discouraged, he wanted to see why he could not give up. And when he scored small victories, he wanted to see who would benefit.

Afterward, they sat around the dining table, a group of people who had never met one another but were distantly related and genetically linked by the mutation in their bloodline. It was their second day in Phoenix, and the experience was sobering, bonding; the potential history of the moment loomed large in the atmosphere.

"This is so hard," someone said at last. "Half the people at this table are going to die from Alzheimer's disease."

Thinking about it a year later, answering the question he had been asked countless times, Tariot said: "So why are we inspired? How could you *not* be?"

Obtaining visas for natives of Colombia—a country the US State Department eyes with suspicion—was just one of the myriad complications facing the Banner team as they prepared for drug trials. For starters, they needed to find a drug, and it had to be the right one. The *paisa* had been waiting generations for the medical field to advance to the point where it might start offering answers instead of asking questions; the doctors were mindful of that burden and of the narrow window of opportunity that had opened to them. But they also had to run trial participants through a battery of tests to determine their baseline status on key biomarkers. In short, the doctors wanted to observe the amount of amyloid they had in their brains, if any, as well as other indicators of how the body was manifesting the disease. These results would serve as a basis of comparison later, after the *paisa* received the drug injections, to determine whether the medication had had any effect.

While the research team searched for the right drug to test, the *paisa* came to Phoenix to get their baseline PET scan readings. It was critical for the team to know when amyloid began depositing in the brain, so they could identify the youngest possible age for inclusion in the trial.

What they found shocked them.

Doctors normally describe the early stages of Alzheimer's as mild cognitive impairment. If you start to forget appointments or can't remember the steps involved in completing a task, mild cognitive impairment (or MCI, for short) is likely the term that will appear in your medical records. In people who carry the mutation, this description might apply in their midforties, a startlingly young age. But what the Colombians' brain scans revealed was proof of what people like Paul Aisen had been saying: that even a decade before that, long before a person's behavior is at all affected, the brain is changing in ways too subtle to appear to the outside world.

In the *paisa*, amyloid began increasing at the age of twenty-eight and appeared to plateau at the age of thirty-six, but mild cognitive impairment wasn't noticeable until an average age of forty-five. A diagnosis of dementia happened at around fifty-one. This is consistent with the idea that the tau cascade happens after the amyloid plateaus and is associated with the worsening symptoms.

Since the subjects were all from the same family, and lived in the same part of the world, the number of biological and environmental variables was much smaller than, say, with the DIAN test subjects, who came from all over the world and could have been affected by any of the three known mutations.

"This is a special population," Reiman said of the *paisa*. "The power we have to see things with this one mutation, it's amazing."

What's more, he saw similar changes occurring in the brains of people with the more common ApoE4 genetic risk for Alzheimer's, the type that both the Banner group and Aisen and Sperling were studying. The brain of a thirty-year-old ApoE4 carrier, for example, wasn't metabolizing glucose as quickly. Some middle-aged people experienced brain shrinkage. And the higher the genetic risk for Alzheimer's among people in their early sixties, the more amyloid there was in their brains. Those results convinced Reiman that the disease was similar enough in both populations that anything that succeeded with mutation carriers would also benefit people who suffered from the more common form of the disease. They appeared to be on the right track.

As the Colombians rotated into Phoenix for their baseline scans, Tariot and Reiman invited fellow Alzheimer's researchers from around the country—who were not directly involved in their efforts—to attend meetings so they could outline their idea for a drug trial. There, they made their pitch: If preventing amyloid accumulation before the tau cascade began would halt Alzheimer's, this was their chance to prove it.

To spread out the risk pool—so the *paisa* would not become the sole guinea pigs for an American pharmaceutical company—the study would include a handful of Americans under the auspices of the FDA; in Colombia, that country's equivalent—Instituto Nacional de Vigilancia de Medicamentos y Alimentos, or INVIMA—would regulate participation.

As is often the case with scientists, the researchers' meeting hosted by Banner yielded no complete consensus; some said the Banner team was onto something important, some said they were setting themselves up for an impossible task. On the final day, each person in the room briefly summarized his or her opinion. The last doctor to speak was Francisco Lopera. His closing words were: "Remember that our families are waiting for you."

While members of the Banner team looked for an anti-amyloid drug to test on the Colombians, they had to court the pharmaceutical companies carefully and ask unexpected questions. If the companies had ulterior motives, they wanted to suss those out.

One of those questions was: Why are you willing to put this asset at risk? The very idea of spending millions to create a pharmaceutical and offer it to a wholly unique population was riddled with business risk factors that many companies would consider unacceptable. The drug could unveil complications unique to the *paisa* or the PS1 mutation that might doom its prospects for the general population; it could prove to be unsafe; there were no guarantees the FDA would approve it. If they did, competitors looking at the study's results could potentially use the information to reverse-engineer the drug and steal it. And to date, not a single Alzheimer's drug had succeeded in budging the disease.

Tariot and Reiman sat through this dialogue with more than twenty companies. One of them was Genentech, a San Francisco–based biotechnology company that was acquired by Roche in 2009. Founded in the mid-1970s, the company was holding among its assets crenezumab, the antibody designed to latch on to—and clear—amyloid in the brain.

Crenezumab began its life as a molecule invented by scientists at a Swiss biotechnology company, AC Immune. The company was founded in 2003 on a shoestring budget of $3 million by Andrea Pfeifer, a German scientist who hoped to help address unsolved public health issues. AC Immune sold its molecule to Genentech, which developed it into crenezumab.

Months after the sale, Pfeifer learned about the Banner Institute's plans to conduct Alzheimer's research in Colombia. She was so touched by the situation that she considered fund-raising for the cause, never knowing that it was her own invention that would be used in the trial. The Banner team selected crenezumab in 2012. When the AC Immune team learned of the connection, it somehow seemed to validate the nine years of work they'd spent developing the molecule. The entire group erupted in celebration.

When asked if she thought crenezumab will work, Andrea Pfeifer spoke with certainty. Even in high doses, crenezumab didn't carry the side effect that had sidelined other anti-amyloid antibodies—vasogenic edema, an abnormal accumulation of fluid and tiny hemorrhages in the brain. (This complication was also known as ARIA, or amyloid-related imaging abnormalities.) Crenezumab binded to amyloid beta by recognizing when the protein's composition had changed from healthy to toxic. Certain that the drug would sweep amyloid beta from the body, Pfeifer said the real question was: Is that the root cause of Alzheimer's disease?

"I have a strong belief that it will do something, yes," she said. To her, the only variables were related to optimization: What is the best timing? What is the best dose? What is the best schedule?

And just in case amyloid beta wasn't the only target, AC Immune was also working on the development of an anti-tau molecule.

If her efforts achieved the positive results that had so far eluded other treatments, "honestly speaking, for me, it would be the fulfillment of a dream," she said. "We do hope that our molecule—*any* molecule—will change the destination of these people and their families."

Because of the potential risk of working in Colombia, the company's leaders wanted to travel to the country to see the setup for themselves. The team met Lopera, saw the hospital where he worked, examined the satellite locations where participants would travel for their experimental shots. When they returned, they were convinced.

Carole Ho, the neurologist who oversaw Genentech's early clinical development group, recalled driving through the precarious mountain roads of the Andes, where Lopera once traveled by horseback, and realizing the dedication that had driven him for twenty-five years. As she explained it, Genentech's motivation came from the belief that science was on its side.

"It's that deep scientific conviction," she said. "You should believe in the science, and follow the science."

Identifying amyloid as a certain target would count as a significant victory, even if Genentech didn't reap a huge profit, she said. And she was impressed by how homogenous the study subjects were in Colombia.

"It's in a lot of ways the most controlled clinical trial that you could embark on," Ho said.

To Tariot, the company's decision was nothing short of brilliant. Genentech was a latecomer in the market; by the time the company had teamed up with Banner, several competitors were already in late-stage human trials and attracting an enviable amount of publicity.

"Here's Genentech behind a bunch of big boys in terms of immunization therapies for Alzheimer's disease. And they say, 'Well, how could we do better?' And so they end up with a compound that is more potent, targets more broadly, and is apparently free of this potentially important liability," Tariot said. "I suppose a subtext to all of this is: Is it good to be first, or not? This might be an example of it being good *not* to be first."

As the pieces for the Colombia clinical trial fell into place, Reiman and Tariot's plan began to take shape. While they tried to come up with a strategy for winning regulatory approval from the FDA, they sought the advice of Jur Strobos, the former director of the policy research staff in the FDA commissioner's office. Strobos offered encouragement: Their idea was not impossible, he said. The problem was that most clinical trials for central nervous system disorders just didn't fit into the FDA's traditional avenues for approval. But they could find clues for success in other diseases, including HIV, cancer, and heart disease.

What the Banner team should not do, Strobos warned, was to approach the FDA out of the blue with a plan and a drug, begging for approval. Instead, he suggested a compelling rationale for an unorthodox plan that was well designed, with evidence backing up the reasons for each element of the design, and the consensus of the rest of the Alzheimer's field indicating that this nontraditional approach was the way to go.

Scientists, patient advocacy groups, the NIH, international experts—all had to agree that the trials weren't some half-baked idea. Check, check, and check. All systems were go; now it was time to see what crenezumab would do. Participants would get two shots under the skin every two weeks for the duration of the study. Because it was a trial aimed at proving a clinical benefit—that the drug would help people think and function better, as observed in their behavior—it was a tall order, and some might have called what Banner planned to do nothing short of grandiose.

It was an adjective Reiman had learned to embrace, even as he acknowledged the uncertainty of what they were about to do.

"We think we have a chance. No guarantee—we think we have a *chance*," he said. "Speaking of grandiose, our mission is to end Alzheimer's without losing a generation."

On December 23, 2013, the Banner Alzheimer's Institute announced that the team had officially administered the first doses of crenezumab to participants in Colombia. For Francisco Lopera, it was a moment he had waited more than three decades to witness.

The clinical trial in the *paisa* families will be formally completed in 2020, the point at which the results can be considered truly definitive, says Ken Kosik, Lopera's longtime research partner. In a context where seven years may see the death of several of the study's subjects with no further answers, that thought gives both Kosik and Lopera pause; they have walked, quite literally, down so many crumbling roads just to reach this point, largely without recognition, often facing hostility from the world outside their science.

"We don't have time to do all the research we can do," Lopera said ruefully. "We will not have time. Because we are using all our lives studying these families, and *now* we are starting."

Kosik agreed: "We are now old, and this project is just beginning."

Twenty-Six

LIKE FATHER, LIKE SON

OVER THE YEARS, oil companies moved Dean's best friend, Monte Olsen, around the country; such was the nature of the job. One summer, he was asked to consider Dubai. After consulting with Dean and another friend, he turned it down. But then the company came back and offered him Australia.

At first, Monte declined again; his youngest son was just about to graduate from high school, and he didn't want to uproot the boy. The next time he was asked, he accepted. But Dean was very much on his mind. In the first year and a half he was in Australia, Monte came back to the United States to attend five funerals. Despite his best attempts to ignore Dean's fate, he was acutely aware of how little time he might have left with the man he considered his brother. And like Becky Vork, who was so devoted to Dean's Wisconsin cousin, Colleen, Monte's true friendship was a blessing.

They struck a bargain: When the time came where he could not live on his own, Dean would move in with Monte, who would hire caregivers. It's a promise that husbands, wives, children, siblings, and friends have made to thousands of Alzheimer's patients over the years. It was, in fact, the promise that Sharon DeMoe made to Jerry. The reality of keeping that promise is something most people can't anticipate: The hallucinations. The fear. The wandering. The repetition. It means mopping up accidents, hiding valuables, hiding weapons—or objects that can be used as weapons. When the patient is only middle-aged, it means physically restraining the sudden violence of someone who is still strong, who has no boundaries—who might strike out at children, at old people, at women, even if that would have been unthinkable before the disease. In fact, Alzheimer's patients are often most violent toward the people they love best, as Gail had learned.

Monte, veteran of the decline of Brian and Doug—as well as ill health in members of his own family—knew better than most what he was promising to do. His plan did not change.

He never thought it was a good idea for Dean to learn his genetic status, believing somehow that the knowledge would create the reality. If he never found out, he never needed to believe he had it. But since Dean went ahead and did it anyway, Monte promised to remain loyal.

"I don't want to accept it, and Dean knows that," he said. Still, Dean talked to him about it. He told him about the research he was doing in Pittsburgh. He talked about his hope of helping someone else. He became, in Monte's eyes, a hero.

"This gene will kick in when it decides to kick in," Monte said. "I hope something doesn't happen to him before it happens to me."

Living in Tioga during the workweek meant that Dean had ample opportunity to spend time with Doug. But while other family members made time to go to the nursing home, or to take him on outings, Dean became increasingly distant. Watching his brother struggle was just another reminder of what he was up against. He visited, but only every so often; and when he did, it ruined his day.

The marked difference between Doug's decline and Dean's relative stability illustrated another puzzle that Alzheimer's research had long sought to explain: what caused such a wide gap in the timing of symptoms among family members? Doug and Dean were only one example; there were also Pat and her oldest daughter, Dawn, both of whom had been in nursing homes at the same time; and Julia Tatro Noonan and her identical twin sister, Agnes, whose symptoms—and deaths—happened about a decade apart.

In Santa Barbara, California, Ken Kosik's university research team thought it might have one answer. Matthew Lalli, a postdoctoral student working in Kosik's lab, sequenced the genomes of one hundred members of the *paisa* family. In people whose onset was delayed, they found a gene variant preventing the accumulation of a protein called eotaxin, the levels of which rise as people age. The team theorized that eotaxin levels might affect the age of onset in Alzheimer's, especially since aging is the single highest risk factor for developing the disease among the general population. Unlike epigenetic changes—which do not change the DNA sequence, but do modify DNA in other ways—the eotaxin variant actually changes a nucleotide in the person's DNA.

Kosik's group partnered with another University of California team in San Francisco to measure eotaxin in 150 people with Alzheimer's or dementia. People with the same gene variant suppressing that protein that had been found in the *paisa* also had a modest delay in the onset of their Alzheimer's disease.

Kosik said the variant occurs in about 30 percent of the general population, and could potentially be a therapeutic target to delay Alzheimer's. But he cautioned that more research needed to be done in larger studies to evaluate the team's findings.

While Dean's family was also reluctant to accept his fate, they were fully supportive of his efforts to combat the disease. His son, Tyler DeMoe, made his first trip to Pittsburgh when he turned twenty-two, and he was eager to participate in the research for which his father and extended family had been a mainstay.

When the DIAN study's doctor told him how important his contribution to research could be, Tyler thanked him for the opportunity.

"I am extremely humbled to be a part of these studies, and very excited to think that I may be able to contribute to a breakthrough," the doctor said.

"Yeah, I feel the same," said Tyler. And he meant it.

By the time Tyler sat down that Monday morning to answer questions, he'd already been in Pittsburgh for two days—enough time for him to have a few adventures. The study's protocol required him to bring a partner who could separately answer questions about his cognitive abilities. For patients like his father, or like Sherry DeMoe or his aunt Lori, the study partner offered an objective point of view: whether the patient was struggling more with memory or becoming more agitated, for example. People with Alzheimer's often don't perceive even obvious changes in themselves.

But for Tyler, who had no symptoms and didn't even know whether he carried the gene, the study partner was merely a formality. He chose his friend Zack Brorby, whom he'd known since grade school. They were neighbors, and now they were students together at the University of North Dakota.

The day they landed, the boys headed down to the city's South Side to find a sports bar so they could watch some college basketball. They settled on a place called Mario's and began talking to other patrons at the bar, who asked where they were from.

"North Dakota," Tyler answered.

"What brings you to Pittsburgh?"

"It's a long story," he said at first. After a few beers, he switched his answer to: "I'm participating in some research."

"You're going to be a doctor?"

"No, I'm gonna get studied," he said.

Some people were genuinely interested; one guy shook Tyler's hand. People were intrigued when he said the research was for Alzheimer's disease; he barely looked old enough to be allowed in the bar.

Tyler was used to it. Most of his friends knew about his family's history, but they didn't necessarily understand the full implications. For example, a lot of them didn't know Alzheimer's was fatal, and he had to explain it to them.

"I'm like, 'Yeah, man.' It's intense when you talk about it," he said. But he refused to let it scare him. He was thoughtful about the disease, but it didn't consume him. He believed in living life as fully as possible. Tyler was ever his father's son.

He moved through the battery of tests at Pittsburgh like a pro. Every morning he woke up, and he and Zack left the hotel, passing the same bum hanging out in front of a gas station across the street; Tyler gave him a cigarette every day. He charmed everyone on the staff at the Alzheimer's center, answering all the screening questions with endearing honesty. When a social worker asked him if he abused alcohol, remembering Mario's, he said, "Well, I might have abused it the first night I was here."

The MRI, which had spooked his aunt Karla and driven other family members crazy with its loud noise, lulled him right to sleep. He was fascinated by the PiB that was created down the hall. Even the PET scan, which required him to have a soft plastic mask molded to his face to keep his head still while the scanner operated, didn't faze him; he asked if he could keep the mask afterward as a souvenir.

"I might wear this out tonight," he said.

After talking to a genetic counselor, he decided that, at least for now, he didn't want to learn his genetic status. There would be plenty of time for that in the future, if he needed to know.

Twenty-Seven

ALL THE CARDS ARE ON THE TABLE

THE SAME SUMMER that Tyler visited the University of Pittsburgh for the first time, his aunt Lori McIntyre came back for her annual trip.

Seated at a conference table, surrounded by an elite team of researchers, Lori was as relaxed as she'd ever been. She showed off her new T-shirt, which bore the ironic motto *Where am I?* The staff who hadn't met her on previous visits smiled nervously at her joke, not knowing her well enough to understand her sense of humor.

She and Steve had traveled the previous day from Wyoming, where they had moved in 2010. In Idaho, when Lori's memory started to falter, they decided their gypsy days were done. Steve requested a job transfer to a location where his wife could be among their closest friends.

"I threw my hardship card on the table and wrote a letter to the [Union Pacific] officials in Omaha. They all knew about Lori's disease," Steve recalled. "Wrote 'em a letter and said if a job opening ever comes up in southeast Wyoming, western Nebraska, I'd like to be considered. Five days later, they called me and said, 'You're going to Wyoming.' "

They settled once again in Laramie, in a house that looked out toward the majestic peaks of the Snowy Range, just a few miles from their youngest daughter, Chelsey, and their only grandchild, Claire.

When they went to Pittsburgh for the research, they were accompanied by Lori's friend from Nebraska, Robin Tjosvold, the nurse who had helped with Chelsey's birth. Ironically, though many people considered Lori one of their closest friends, very few stuck around when her disease began to show. Of the three women who had vowed to serve as surrogate mothers to Lori's daughters, only Robin had lived up to her word. The others made the occasional effort with a phone call or a letter. They'd been sidelined by their own lives, their own health concerns, their own grandbabies. It was Robin who drove out to visit Steve and Lori, who danced with her, comforted her, still laughed with her, still made her feel human.

On Mother's Day, Lori's three daughters gave Robin a card, too: Its cover showed two old ladies dancing, with the caption *They dance like they're the only two who can hear the music.* It suited them perfectly.

As a nurse, Robin Tjosvold had worked in settings ranging from delivery rooms to hospices. She gave the McIntyre girls a single piece of advice about their mother: Do whatever you can now to avoid regret later. She knew what she was talking about. When she was eighteen years old, a brain tumor had killed her father. She remembered how, in the end stages of his life, he'd asked her, "Do you feel sorry for me?"

Even at that age, she knew the answer: No. She felt sorry for herself, and for the rest of her family, for the time they would not have with him.

As she sat in a waiting room at the Alzheimer's Disease Research Center at the University of Pittsburgh, Robin struggled to maintain her composure. Her experiences as a nurse had taught her to be detached, pragmatic. But this disease was so complex, and her friend—once so strong—was now so vulnerable. Robin was touched by how tirelessly Bill Klunk and his team were working. As she thought about the massive scope of the problem they were trying to solve, she was overcome.

Finished with another round of tests, Lori walked into the waiting room and saw her friend's tears. She wrapped her arms around Robin and let her weep. When it was over, Lori said, "I just forget so often what it must be like for you to see all of this."

Robin witnessed how uncharacteristically dependent Lori had grown on Steve, and how courageously he fought to take care of her even as she chafed at his efforts. Like most people who knew them, Robin was surprised at his dedication. Years earlier, before her symptoms worsened, Lori had confessed some of her fears about Steve to her friend.

"I'm having a hard time trusting that he will be there for me. How's he going to do this?" she asked. Her independence was one of the qualities he had always loved about her. When that was gone, would he view her differently? Would he still love her, or would she become a burden?

"All you can do is trust that he keeps his promises," Robin said.

Indeed, Steve had never been a patient man, never been a touchy-feely kind of guy. So seeing him take such painstaking care of Lori—fighting to hold his tongue when she complained that he hadn't told her something, waiting out her inevitable meltdowns, reestablishing roots in Laramie to keep from disrupting her world—made his daughters love him even more fiercely. The Tioga relatives didn't understand why he would not bring her back to North Dakota; but he knew, and their girls did, too: In her right mind, she would not have wanted that. He would fight to avoid that move for as long as possible.

"He's taking the whole thing like a champ. I never, in a million years, would have guessed that," his daughter Jessica said.

Jessica was Lori and Steve's oldest child, but the one most like her father in personality—"we're both kind of hard-asses" is how she put it. She asked him whether he might need a support group. No, he said; I'm not going to do that stuff. He confided to his friends, his family. Alzheimer's support groups were for people his parents' age. He thought for a while about keeping a journal, but he worried that he'd wind up writing something in it that he'd later regret, a permanent record of temporary frustration. So he kept his troubles mostly to himself.

Immediately after her mother's diagnosis, Jessica searched for a geneticist in Denver, where she was living, who could test her for the mutation. It was a more difficult task than she had anticipated.

"Straight up, a ton of people said, 'No. I don't feel comfortable doing that,'" she said. It was the same dilemma with which Ken Kosik and Francisco Lopera wrestled in Colombia: Without a meaningful treatment for Alzheimer's, such knowledge can be treacherous, triggering dangerous reactions in people, and many doctors were reluctant to be part of that process.

Finally, a doctor at the University of Colorado agreed to run a blood test if she saw a genetic counselor. Jessica paid for the entire ordeal out of her own pocket. It was beyond her means financially, but emotionally, she couldn't afford not knowing. Jessica believed in confronting problems head-on, and ambiguity would have driven her crazy. For twelve weeks, she waited on tenterhooks, until the result came back: She was free of the mutation.

Nobody in her immediate family had ever died or been terminally ill before, at least not anyone she knew. She wanted to do something; sitting back and passively allowing the situation to unfold was simply not in her psyche. She began volunteering with the Alzheimer's Association, dragging her friends along to events. She sat on a panel at the association's annual educational seminar and spoke about her family. At one fund-raiser, she delivered an eight-minute speech in front of an audience of about five hundred people, curing in one fell swoop her fear of public speaking. She made plans, with her sister Robin, to write a play about Lori's life.

But as her mother declined, so did Jessica's seemingly limitless kinetic energy.

"She's always been everybody's mom. She's in charge; she's the boss," Jessica said. It was unfathomable that her mother, the strongest, most fun-loving person she knew, would no longer be able to organize camping trips, fix electrical outlets, or listen to everyone's problems. She was the center of their world.

To Steve, it wasn't just the gradual loss of his wife that stung. It was also the gradual disappearance of their friends.

In Pittsburgh, he noticed how Lori seemed to come alive with each new person she met; at home, she was bored to tears, unable to work. Steve had built her a craft room alongside his workshop, but it was filled with yarn, needlepoint, and scrapbooking supplies that she no longer knew how to use. When she tried to read a novel, the bookmark stayed stuck in the same spot for months, or went backward as she forgot earlier chapters and retraced her steps. She maintained subscriptions to *Reader's Digest* and *Good Housekeeping*, but a single magazine looked like she had tried to read it for years. Once, she'd been able to whip through jigsaw puzzles in a day or two; now she had one that she had started at Christmas, and months later, she was still toiling away.

Ironically, the woman who had mastered the art of multitasking had been relegated to spending most of her day in front of the television, feeding her sweet tooth with the candy that she hoarded. She was unable to remember that she was not hungry.

When their dog died on Christmas Day in 2011, Steve tried to cheer her up with a new Labrador retriever puppy, but Lori became frustrated that the dog responded more to Steve than to her. When he came home after work in the evenings, he found wadded-up tissues, still damp from her tears, squirreled around the house.

Some of her struggles were his alone to witness. When they were getting ready to go to Pittsburgh for a research visit, Steve suggested that they pack the night before, since they were scheduled to depart early the next morning. Lori said she was already packed, but her suitcase was empty, her clothes still hanging in the closet. All their lives, she had been the one who planned out their trips; now Steve tried to fill her shoes. Gently, he guided her into putting her clothes in the suitcase, just as Becky Vork had packed for Lori's cousin Colleen.

Again and again, she asked Steve: *When are we leaving?* With as much patience as he could summon, he answered: *Early tomorrow morning.* Finally, they finished, and Steve went out to his shop, attached to the garage, where he stored his own suitcase. When he returned, he found Lori locking doors, shutting off lights, and gathering her cell phone charger.

"Honey, we're not leaving until the morning," Steve said quietly.

She looked at him, crestfallen.

"You confused me!" she said accusingly. And then she stormed into their bedroom and began sobbing.

Steve gave her a moment, then went in to console her.

"Maybe I didn't tell ya," he said. "We're leaving in the morning." And with that, she seemed to feel better. Just as Gail had told Deb, Steve had to find the patience to be wrong.

Once a week, Lori did still drive to a coffee shop to meet with a group of girlfriends. She looked forward to it all week; it was her one regular diversion, though it would soon have to end since she'd gotten in a car accident with her granddaughter, Claire, in the backseat.

Lori believed she was still managing the household and that her moods hadn't changed much. But Steve told the Pittsburgh research staff otherwise. He had taken over all the cooking. Instead of washing dishes, Lori put them back in the cupboard still dirty, forgetting to load the dishwasher. Instead of doing loads of laundry, she would wash two socks at a time, unable to recognize that the basket wasn't full.

A few months earlier, visiting Gail's house in North Dakota—the very house where she'd grown up—she'd gotten up during the night to use the bathroom and become lost. And she became temperamental with Steve, exploding into outbursts. When they were on trips with extended family, she sometimes took other people's belongings, stashing them in her suitcase.

Her erratic behavior—which mirrored some of the symptoms her late cousin Dawn had exhibited—led Steve to think that the time was drawing near to move Lori into some kind of nursing home. He was still a few years away from retirement, and he worried about her safety at home alone when he was at work.

Occasionally, some of their friends would call or send him an email, checking in to see how Lori was faring. Always, there was the same rote offer.

"I find it kind of funny, or kind of ironic," Steve said. "Every time you talk to somebody—one of your friends—[they say], 'If there's anything you need from us, just let us know.' I'm thinking, 'OK, how about you invite us over for dinner Saturday night?' 'Take Lori to the movies for me on Sunday, will ya?' "

He chuckled wryly. "I want to say that, but . . ." His voice tapered off. "I have a very strong feeling about this myself, and it was told to me once before, and I'm starting to realize it now. When you have a disease like this, you tend to realize who your friends are. 'Cause there's some friends out there who've just kind of basically turned away, shied away. Scared."

But Lori said the silence of her friends was understandable.

"That may be true, but that doesn't bother me, I don't think," she told Steve. "Our family chose to go forward. . . . I think we just kind of, all of us, made a choice to not feel sorry for [ourselves]."

Steve knew his wife had a point. "It's just been accepted that there's a disease that we're gonna have to fight, and we're going to fight it," he said. "We're going to do everything we can to help find a cure. All the cards are on the table. We just need to play those cards going forward."

They were still able, in 2012, to speak to the doctors in Pittsburgh about Lori's plan to continue contributing to the research after her death. Like her brother Brian and her favorite cousin, Dawn, she wanted her brain to be sent to the university, where it would be examined in a separate autopsy—just as her father's had been, and Auguste Deter's a century before.

In fact, Lori wanted all her organs donated to research.

"Is it OK with you?" she asked Steve, as she signed a form verifying her interest in the brain autopsy program.

Steve gave his support, and they agreed on a final plan: "We're gonna have you cremated, and I want to take you to every place that we've ever been, and drop an ash somewhere."

The belief that she could contribute to a solution, possibly her last act of sacrifice as a mother attempting to change the course of her children's lives, drove Lori fiercely through the research at Pitt.

Sitting next to the PET scanner that had just taken detailed images of her shrinking brain, she shivered. The room was chilled for the benefit of the equipment, so nurses brought her blankets they'd warmed to make her more comfortable before her next round of scans, this time to measure how her brain metabolized glucose. She was worn out by the testing, and her brain was struggling to process its disorientation from the trip. Every time she traveled now, she came back permanently worse.

How would you like to be remembered? she was asked.

"That's a hard question," she said. And she gave the most honest answer she could, one that reflected who she'd been, at her core, her entire adult life:

"Happy," she said. "Hopefully, a good mom."

Months after that 2012 visit, during which she underwent her seventh set of brain scans, she and Steve sat on a conference call with Bill Klunk, who reviewed her results along with the neuropsychological tests she had completed.

Before he started, Klunk made small talk with the McIntyres. He'd come to know them as well as anyone ever had, examining the inner workings of their brains but also the interior map of their hearts. He asked about Claire, knowing Lori's granddaughter meant everything to her. Then he moved on to the crux of the conversation.

"As you've told us, each year it's been a little tougher for the memory, right, Lori?" Klunk began.

"Yes," she said.

"And that's exactly what our tests show," he said. In the 2,079 days since she first underwent research at Pitt, Lori's cognitive function had gradually declined. She learned six out of ten words the research team had asked her to remember, but she forgot all of them twenty minutes later. She had difficulty with an executive function test that asked her to draw a clock face. She remained relatively strong in visual-spatial reasoning, a rare bright spot in her report.

"But I think this pretty much tells us the picture, and the picture is consistent with what you told us," Klunk said. "Memory this year isn't as good as it was last year."

The MRI revealed a little more brain shrinkage than a similar scan from

2008, but there were no major changes such as strokes, tumors, or rapid shrinking.

"All in all, there's a little more shrinkage than we expect for your age, so a little more than an average fifty-three-year-old, but not anything tremendous," said Klunk.

Klunk compared her most recent PET scans to the originals six years earlier and said the trend showed a steady decrease in her brain's use of glucose, consistent with a pattern seen in Alzheimer's patients, particularly in the temporal lobes right above Lori's ears. The largest cluster of amyloid was in her striatum, the very center of her brain, which usually occurs much later in late-onset Alzheimer's patients. On the other hand, her brain hadn't accumulated much additional amyloid across the six-year span; rather, the doctors felt that the changes Lori was experiencing were the manifestation of what had already accumulated for twenty years.

That remained consistent with the Baptists' working theory that amyloid levels built up early and triggered the mechanisms responsible for later symptoms. In time, even dedicated Baptists such as Paul Aisen, the neurology professor from the University of Southern California who was investigating Alzheimer's in the general population with Reisa Sperling, would come to believe that tau levels rose steadily throughout the course of the disease, and that tau correlated with the later degrees of impairment that Lori was now experiencing. But PiB only measured amyloid, and there were not yet ways to look at tau to measure its growth and compare it to her behavior.

"That's one of the things we're trying to figure out. It's not just having the gene, it's not just having amyloid in the brain. It's other things [that contribute to Alzheimer's], and we'd like to know what those other things are," Klunk said.

But he added: "Look at it in the perspective of the family: Lori's holding her own really well."

"Thank you," she said.

"Given the genetic issue that she's been forced to fight, you're putting up a pretty good fight, Lori," he added. "So keep fighting. Keep physically active, keep socially active. Keep fighting."

Steve had the idea to stage one more bash for Lori, one more chance for her to embrace and celebrate with her friends before Alzheimer's made that impossible.

It was Labor Day weekend, Lori's fifty-third birthday, and Steve planned a blowout of a party that attracted friends and family from far and near to their house on the outskirts of Laramie. Karla and Gail, who knew they wouldn't be able to make it that weekend, instead drove out earlier that summer for their own visit. Since both were hopeless at directions, they enlisted Tyler to drive them, and they had a wonderful time.

When her birthday finally arrived, Lori held court, radiant in a printed yellow-and-orange top, her eyes alive with the excitement of so many people she loved gathered in a single spot. Brian's son, Yancey, and his wife and newborn baby daughter drove all the way from North Dakota; Lori ran out into the yard to greet them. She was overwhelmed that everyone would come so far, for her, only for her.

Her daughters presented Lori with the birthday gift they had been compiling for months. The idea was as simple as it was profound: contact as many people from Lori's past as possible, and have them write a letter about a memory they had of her, accompanied by a picture.

The McIntyre girls bound the responses into a single album for their mother to thumb through, reading about the woman she had been. There was a friend from her Tioga childhood who remembered how Lori and Karla would dress Jamie up in hair ribbons.

There were pictures of Lori in 1977, about the time of her high-school graduation, sporting her ultra-short hair; a reluctant Lori in a flowered prom dress, about to head to the dance with one of the local boys who had always bored her. There was Lori as a young bride in the dress Steve's mother had sewn, as well as a copy of the 1980 pattern she'd used to make the gown.

Leafing through the pages of collected memories, some of which she could still recall and others which were now missing, Lori read about herself: an interesting, spontaneous woman who had lived her life fully and well, who was simultaneously nurturing and restless, always seeking adventure.

The party spilled out to the driveway, the garage, and Steve's workshop, under the bleached skull of a buffalo he'd once shot. In one corner was a stand overflowing with costumes: hats, feather boas, wigs, and props. Lori, like Gail, loved to clown around in dress-up clothes. Steve set up a discarded plow disk over a propane flame and cooked burritos; he grilled buffalo and elk steak to mouthwatering perfection, too.

Everyone at the party wore cowboy boots. Little Claire, the life of the party, wore a yellow University of Wyoming T-shirt with the motto *Pistol Pete is my homeboy.*

It had been a dry summer, and the view of the mountains on the drive into Laramie had been hazy from the smoke of fires fanned by warm winds. When rain finally rolled in that night, they could see it coming across a vast canvas of sky, terrible and beautiful, supernatural, the clouds godlike in their majesty. Though you could see the rain advancing across the endless horizon, miles in the distance, it still might not fall. In a drought, there were no guarantees.

But the rain did come, in warm, fat drops that spilled out of the sky and into Claire's face and hair when she ran into the night. As they fell, she began dancing, her cowboy boots stomping on the pavement. She threw her arms up in the air, as if to celebrate the arrival of the fire's long-awaited cure.

Twenty-Eight

COMING HOME

FOR ALL THE pragmatism Steve and Lori McIntyre had demonstrated when talking about how they would handle her disease progression, little could have prepared them for their final days living under the same roof.

In the summer of 2013—just a year after Lori appeared to be holding her own in her tests—their middle daughter, Robin McIntyre, moved from Denver back to Laramie to help take care of her mother. During just the past few months, Lori's mood swings had begun to careen in hairpin turns at the slightest triggers. When she disagreed with someone over the phone, she hung up on them. When she was upset, she locked herself in the bathroom. She stopped finding the words to express her thoughts, and she could no longer write them down, either—the art of paper and pencil had disappeared into the constellation of amyloid and tau proteins scattered across her brain.

Steve wanted to hire a caregiver to come to the house when he was working and Robin couldn't stop by. Lori wouldn't hear of it; she didn't need a babysitter, she said. But she began getting up in the middle of the night, trying to figure out where she was. If she decided to go wandering outside when Steve was asleep, she could very easily die. They were in a city by Wyoming standards, but it was a city on the outskirts of a vast mountain range, rich with wildlife; at night, they could hear the coyotes gathering close to their house.

Steve would come home to find that Lori had unloaded all the contents of their refrigerator and freezer. Robin drove to her parents' house to find pans on the stove, with Lori nowhere to be found. Thinking her mother had locked herself in the bathroom again, Robin decided to wait. But after a while, when Lori didn't emerge, she checked the bathroom and discovered that it was empty. Robin began searching; eventually, she found her mother sitting in the craft room next to Steve's workshop. Lori was holding some kind of document, weeping as she read it.

It was Galen DeMoe's autopsy report.

"I didn't know my dad had Alzheimer's," she said.

That fall, when Steve had to be hospitalized with pancreatitis, he reached a decision. He was simply too sick to take care of Lori, and to ensure her safety. There would not be another Christmas together as a family. He would check out of the hospital and cash in his remaining vacation time to take care of the necessary arrangements. Lori was going into assisted living.

At first, he hoped to convince her to check into a new facility in Laramie, where she could see her husband and daughters as often as they liked. It seemed more like a complex of studio apartments than a nursing home, but Lori was not impressed. She referred to it as the "loony bin" and flatly refused to go.

Robin took her mother there under the pretense that Lori could volunteer. It was the kind of work she'd always enjoyed, but Lori refused.

"Nope," she told her daughter, unconvinced. "You're trying to drop me off."

The impasse was as excruciating as it was untenable. Nobody wanted Lori to end up like her father or her uncle Jerry, hauled away against her will, baffled and terrified. In desperation, Robin turned to her aunt Karla, veteran of Brian's and Doug's nursing-home transitions.

It was Karla who first suggested the idea that Lori might be better off in Tioga than in Laramie. Perhaps the memory of her frosty conversation with Steve after Lori's diagnosis was still too fresh; perhaps everyone remembered how desperate Lori had been, in her youth, to leave her hometown. Either way, at first, the idea was not universally embraced. But in time, most of the McIntyres began to see some merit in the plan.

Doug was already at the Tioga nursing home, and the family could persuade her that she was checking in to participate in Alzheimer's research and help her brother do the same. Her siblings, nieces, and nephews could see her every day. Moreover, the home was staffed by townspeople they'd known all their lives. They'd watch out for Lori, just as they did Doug.

Because Lori knew her family was participating in the DIAN study's drug trials, it was a credible story. Karla recommended a two-pronged approach: First, Robin would begin talking about the research with her mother, planting a seed. Later, Karla would call Lori and coincidentally discuss the same research.

It worked, just as Karla had predicted. In their phone call, Karla excitedly told her sister, "They're talking about doing a drug trial in Tioga, and Doug's going to be doing it! We've got to get you in on it."

She embroidered on the lie, telling Lori that Jamie and Dean weren't eligible to participate because they were still working. But if Lori came to Tioga for Thanksgiving, Karla suggested, she could stay a little longer to contribute to the study.

Lori, lifelong nurturer and fearless protector, agreed to come. Steve was out of the hospital, so she felt she could safely leave him behind in Wyoming and help the research, which had always meant so much to her.

"Doug is scared. I'll show him how to do it," she said. She asked Karla to explain to Steve the plan she thought they'd hatched together.

To an outsider, such deceptions might seem dishonest, even cruel; but to the families of Alzheimer's patients, they become a necessity. Truth is just another casualty of the disease. Try to ease an Alzheimer's patient into a nursing home, and you could be confirming the abandonment they most fear. Viewed in that context, a white lie that convinces the person to move to a safer environment with a minimum of commotion can be seen as a kindness. But the guilt can still linger.

"No matter what decisions you make, you never feel good about them. You just get that awful feeling that doesn't go away," Karla said.

But she added: "We're only doing the best we can with what we know."

The day after Thanksgiving 2013, the McIntyre girls walked their mother to the front door of the nursing home in Tioga. Steve had driven Lori up the previous week, then returned home before the holiday. After having been off so long with his illness, he had to return to work, and he didn't want to risk upsetting her even more when he left. He had several invitations for Thanksgiving dinner. The house was so quiet with Lori gone.

It was important to Steve that his wife was comfortable and able to accept where she was. He remembered what one of the Pittsburgh doctors had told him: Although Lori had been valiantly hanging on to her mind for a long time, longer than her younger brother Doug, once the cognitive memory slips gears, it disappears quickly.

"We knew it was coming," Steve said, not sure if he was relieved or depressed. "Now that it's here, we're facing the reality."

Karla and the three McIntyre girls got Lori settled into her new bedroom, decorated with pictures from home. Before Karla left, she and Lori hugged.

"I am so lucky to have my girls and you," Lori said.

"We're gonna get this disease in the end," said Karla. And they cried.

Within weeks, Lori had settled in. But by February, although friends and family visited her daily, Lori's patience for being at the facility had just about disappeared. At one point, she packed her bags and announced that if Steve didn't pick her up, she would walk home.

Bill Klunk volunteered to help. They set up a video conference for Valentine's Day, and Steve came to Tioga to be with her as she watched Klunk on the screen. He told her they were all proud of what she had done for Alzheimer's research, and that while she was now going to live in the care facility permanently, her contributions would continue to help the field. She said she understood, but within half an hour, back in her room, she asked Steve when he was taking her home.

The next day, he took her for a walk around her old neighborhood.

"My mom lives around here somewhere," Lori said.

He brought her back to the facility, told her good-bye, and told her he loved her.

I love you. Thanks for being my partner for 18 years.

—card from Lori to Steve on their eighteenth anniversary

Hope the next 18 are just as pleasurable.

—card from Steve to Lori on their eighteenth anniversary

There is a curious symmetry to Alzheimer's in that it causes people to lose abilities roughly in reverse order to the time in which they acquired them. Executive functions disappear first; it's the primary skills—walking, talking, and ultimately breathing and swallowing—that vanish at the very end. Lori's life followed that arc, beginning and ending in the North Dakota town that forged her, repelled her, sent her hurtling into a life of exploration as untamed as her heart. As the path of that life began orbiting back to the small, to the familiar, she was pulled almost gravitationally toward Tioga.

And so it came to be that Lori DeMoe, who couldn't wait to leave, came full circle in returning to her hometown.

Twenty-Nine

PALPABLE MOMENTUM

EIGHTEEN MONTHS BEFORE Lori moved into a nursing home, in May 2012, the Obama administration named Alzheimer's disease "one of the most feared health conditions" and released a battle plan that set a lofty goal: finding a way to prevent and effectively treat the disease by 2025. With the announcement came a promise of $50 million in new research funding for fiscal year 2012, and another $80 million in fiscal year 2013.

It was a signal that at last, the disease had captured the public's attention. The US economy simply could not bear the anticipated burden of caring for a generation of baby boomers afflicted with a prolonged, inconvenient, messy illness for which there was no meaningful treatment.

The first-ever National Alzheimer's Plan enjoyed bipartisan support. Goals included finding any new genes that either predicted or protected against the disease. The plan also sought to boost enrollment in clinical trials and research, especially trials for the most promising drug interventions.

"The question is, can we really beat this disease?" said Randy Bateman, head of the drug trials for the Dominantly Inherited Alzheimer's Network, in which the DeMoes were enrolled. "I expect real, dramatic changes in research and development will occur. . . . There will be a palpable change in what can be accomplished, and what will be accomplished."

Though the price tag might seem high, advocates argued that it was money well spent: Science had won important battles in other diseases when money was invested in finding a cure. Prior to the creating of an Alzheimer's plan, the National Institutes of Health spent about $500 million annually on Alzheimer's, which affects 5.2 million Americans. By comparison, the NIH spends $3 billion a year on AIDS research, which affects about one-fifth as many people. Cancer research gets $6 billion, and heart disease gets $4 billion—and in all cases, that research has translated into better treatments.

Not funding Alzheimer's research could actually prove to be more expensive to the public. The Alzheimer's Association reported that in 2013, the direct costs to society of caring for Alzheimer's patients would total about $203 billion, a figure that includes $142 billion in Medicare and Medicaid costs.

By 2025, the target date for a prevention, the number of Americans sixty-five and older with Alzheimer's was expected to hit 7.1 million, and by 2050, it was projected to mushroom to 13.8 million, at a cost of $1.2 trillion. Economically, Alzheimer's had the potential to deliver a crippling blow.

Bateman's DIAN trial won a significant vote of confidence when the Alzheimer's Association awarded the trial nearly $4.2 million in the form of a four-year research grant, the largest the association had ever bestowed. In October 2011, ten competing pharmaceutical and biotech companies formally agreed to jointly support the study with time, expertise, and money, even if their drugs were not those selected for the trials. That cooperation would prove to be the buoy that kept DIAN afloat when the faltering US economy caused the political climate to sour in 2012.

"If you had asked me five years ago what the odds were that different pharma companies would sign the same agreement to get these prevention trials going, I would have said it would be hard to get one to do it. I would have said it is impossible to get ten to do it. To me, that is a huge step," Bateman told a reporter at the time. The companies were finally on board with an approach that, while unorthodox, could solve a crucial problem when competition failed to make an impact.

At a pretrial meeting, consultants told Bateman's team that they'd never seen a more positive start to a clinical drug trial: It was well designed, and despite its complexities, all stakeholders had been remarkably supportive; it seemed as though everything was going to go off without a hitch. Bateman was riding high.

But the DIAN team had a significant scare at the end of 2012, when a combination of automatic tax hikes and deep spending cuts threatened to take effect on New Year's Eve if Congress could not reach some sort of budget compromise. At stake was their NIH funding, which was crucial to their work.

On New Year's Eve, while lawmakers in Washington debated, the DIAN clinical trial officially enrolled its first patient. In the wee hours of New Year's Day, Congress reached the compromise that delayed the automatic cuts until March; by then, Randy Bateman's team had enrolled ten participants and begun administering drugs; sites in Australia and Europe continued to move through their startup phases, and in North Dakota, Karla was urging Dean to enter the drug trial at DIAN's headquarters in St. Louis instead of waiting for Pittsburgh to get online, a delay she feared would push him past his window of eligibility. Deb concurred, and Dean agreed to make the phone call.

Anticipating that the NIH would not be able to pay for even the majority of the study, Bateman created a backup plan. He began lining up private funding to close the gap created by the government shortfall. He knew there were supporters at the NIH who were pulling together the last of their remaining discretionary funds to try to at least partially fund the trial, but he recognized that it was a political battle. There was, after all, never a shortage of diseases that needed research dollars.

The notion that science could be so close to finding a meaningful treatment, yet lose its grasp to political bickering, frustrated him deeply.

"We've had Alzheimer's for at least five thousand years, likely for as long as humans have been on the planet," with no way to treat it, Bateman said. Now that they had a real shot at a breakthrough, success would require a monumental commitment from everybody involved.

And if competing drug companies and scientists could find a way to do it, he hoped politicians could, too.

Dr. Francis Collins, director of the NIH at the time of the budget cuts, announced that he was designating $40 million from his director's budget for fiscal year 2013 to Alzheimer's research. He also said the sense of momentum was "palpable."

"It would be fair to say that eight to ten years ago, people were feeling a little frustrated, a little discouraged. It seemed as if progress was really slow," he said. "Now, researchers in this field are energized because of a variety of new things that have come along, many of them in just the last couple of years, that give us a sense that we're finally beginning to get a real handle on what causes this disease and what to do about it."

Thirty

TO THE MOON AND BACK

Take up our quarrel with the foe:
To you from failing hands we throw
The torch; be yours to hold it high.
—John McCrae, "In Flanders Fields"

AS MEMORIAL DAY weekend unfurled in 2013, Gail was beginning to feel more like her old self. On Friday evening, she spoke with her sister, Bobbie, for more than an hour. At times, the sisters had had their differences—as many do—but that evening, they chatted deep into the night.

Only a few weeks earlier, Bobbie had surprised Gail with a gift of $5,000, and her timing couldn't have been more opportune. Gail was generous to a fault, and had never been a materially wealthy woman. Gail's health problems were nagging her: a bad back, aches and pains, and as had been the case in recent years, her heart. The medications were expensive, so Karla had added calling Medicaid to her to-do list.

All her life, Gail had been a lively, upbeat woman. Now seventy-eight, she remained the family's rock, a physically tiny woman whose courage and good humor had anchored them through decades of storms.

But that unity had come at a tremendous cost to the woman who provided it. Though antidepressants helped somewhat, those who knew her well—particularly Karla, her confidante—knew that she carefully hid a grief of immeasurable depth. It was Karla who remembered, when many did not, the nervous breakdowns that had landed Gail in the hospital.

Even though Dean stayed with Gail during his workweek, Karla worried that the big house was just too much for her. The cleaning lady Karla had hired was diagnosed with cancer and quit; they had trouble finding a replacement, because nobody wanted to work for such low wages when they could earn better money in the oil fields. But Gail would not hear of moving. She had a community in Tioga, and her house was still a family gathering point, still a place where her friends dropped in.

On Saturday, Dean headed to the other side of the state to spend the long holiday weekend at home with Deb and his kids, a rare respite from the relentless work of the oil fields. As he was leaving the house, Gail asked her son to bring her back one of the ice cream cakes she loved.

Dean shook his head; his house was in Thompson, close to three hundred miles away. How his mother expected him to bring her back an ice cream cake was beyond him. In a cooler, she suggested; he banged out the door and hopped into his truck to make the long drive.

Gail offered to babysit six-year-old Savannah overnight for Chelsey and Jamie that weekend, an unusually busy one for the young family. Gail suggested that she could keep Savannah until after church on Sunday morning. Chelsey hesitated.

"It's not that I didn't really want her to be with Gail, it's just that I was afraid something could happen," she recalled later. Gail's heart was acting up, and Chelsey was always nagging her to keep her nitroglycerin pills with her at all times, to derail a heart attack if she felt one coming on.

But Savannah did love her grandma. And Tioga was a small town, with Gail one of its most beloved residents. If anything happened, neighbors in any direction would drop everything and come to help. Chelsey decided on a compromise, figuring she'd retrieve Savannah late Saturday night.

She pulled in front of the DeMoe family house on North Hanson Street around midnight. Gail—always a night owl—was wide awake. She urged Chelsey not to disturb Savannah, who had fallen asleep. Instead, the two women—separated by five decades in age but united by a sincere mutual affection—sat up and talked until the wee hours of the morning. They discussed old boyfriends, like girls at a slumber party. Gail pulled out her collection of love letters and they read them and laughed. And then, for no real reason, she tied one of her own baby shoes to a heart-shaped wreath on the wall.

At 2:00 a.m., Chelsey—who was tired from her drive—checked on Savannah.

"Just leave her, just leave her," Gail insisted. They'd meet after church the next morning and put flowers on Moe's and Brian's graves for Memorial Day. Chelsey already had the fresh blooms in her car. Gail chattered away as she followed Chelsey out the door.

"Make sure you call me!" she reminded Chelsey.

"It was the best I've seen her," Chelsey said. "She was just so happy, and so full of energy."

The next morning, Gail swapped out her purse before heading to St. Thomas Catholic Church, Savannah in tow. A bit of a wild streak had always run through Gail. She looked like the sweet Midwestern grandma that she was, who crocheted beautifully and baked the most heavenly dinner rolls from scratch. But she also liked to sip her Colorado Bulldogs on the back porch, and she devoured *Fifty Shades of Grey*, the racy novel that was titillating women all across the country. She told bawdy jokes with the best of them; but she was still, in her heart, a deeply faithful person. A half century of profound loss had not changed that.

Halfway through that Sunday's service, during the offering, Gail told Savannah she wasn't well. They ducked outside to avoid disturbing the other worshippers.

Gail took a few steps onto the concrete sidewalk outside the tan brick building and collapsed. Savannah, who had been dawdling with a book, dropped it and ran to her. Her grandma's eyes looked enormous. Gail struggled to form words.

Savannah screamed. She ran back to the church door, but it was too heavy for her to open. She pounded her small fists against it, crying for help, but the music from the organ drowned her out. She ran to the curb in front of the church, trying to flag somebody down. A man in a truck drove past, slowed down, then kept going. Savannah started screaming and beating on the door again. Finally, it opened.

Chelsey Determan was at home, in her laundry room, when she heard someone at her door. By the time she got there to open it, whoever had knocked was gone; she leaned out and saw a woman tearing off around the corner.

Then her phone rang. It was Gail's longtime friend Pat Branesky. Gail had fallen at church, she said; she was at the hospital, and the situation was not looking good.

Jamie took off immediately for the hospital; Chelsey went to find Savannah at the church.

The little girl was terrified. Chelsey soothed her daughter and took her to her parents' house, then left to meet Jamie at the hospital.

"Grandma wouldn't talk to me," Savannah said.

When the word spread to Karla through friends from Tioga, she immediately called the hospital.

"I need to tell you that your mother didn't make it," the doctor said gently. "She passed away before she even got to the hospital."

When she hung up the phone, Karla called Dean.

"Mom's gone," Karla said, and they both started crying.

An hour later, Dean DeMoe was back on the road to Tioga, his little dog, Rookie, by his side. A hundred thoughts immediately converged on his brain: He should have been there. He shouldn't have come back home.

Dean never paid much attention to the fact that he had Alzheimer's disease; it was as though he believed, as his friend Monte had suggested, that he could ignore it into submission. But he did carry a healthy fear of something bad happening to his mother in his absence, and now that fear had been realized: She had suffered a major heart attack. He wished he had just brought her the ice cream cake, like she'd wanted.

When Lori arrived at her mother's house, Deb was the first person she saw. She walked in and gave her sister-in-law a big bear hug, then said: "Hey, where's Mom?"

Over Lori's shoulder, Deb mouthed to Steve: *You didn't tell her?*

Steve nodded his head vehemently: Yes; he'd told her. He'd told her fifteen times. Karla had also told her, over the phone. She couldn't remember. Before they got to the house, she'd tried to buy a postcard for her mother.

Deb pulled herself together, willing herself into teacher mode.

"She's not here," she told Lori.

"Where is she?"

"Well, she . . . died," Deb said finally.

Shocked, Lori looked around. Tears filled her eyes. The grief was fresh, the devastation new; the disconnect between what was unfolding and what she remembered was overwhelming.

"Somebody talk to me!" she said, frustrated.

Gently, Karla led her sister to Gail's bedroom at the back of the house and closed the door. As they had often done in times of crisis, the two sisters lay down on their mother's bed, and once more, Karla explained.

Throughout Memorial Day weekend, the family reunited, streaming into Tioga from all points. With few other places to stay, most of them ended up sleeping at Gail's. For a week, it seemed like old times: People crowded end to end in the DeMoe homestead, drinking, laughing, and occasionally squabbling.

On the sidewalk outside the church, on the spot where Gail had fallen, her grandchildren and great-grandchildren picked dandelions and molded them into a heart. Then children and adults alike gathered in a circle, holding hands, heads bowed, to say a prayer.

The rain that had been threatening to fall all weekend ripped loose from the sky, washing down in unforgiving sheets. The day of the funeral, mourners carried umbrellas out to the cemetery and huddled under a tent erected near the gravesite, but the rain pelted them sideways: a soaking, sheeting, relentless rain. Gail's grandsons bore her casket to the grave, along with a family friend and Kassie's husband, Frankie Rose.

They buried Gail, in the rain, alongside her husband and firstborn son.

The family stayed to sort out loose ends and mourn their matriarch as only the DeMoes would. As they divided up her clothes for people to take home or donate, somebody got the idea to start wearing them. Before long, the entire family—men and women alike—were swathed in Gail's wardrobe; Dean wrapped himself in a multicolored scarf, Lori donned a red floppy hat, Lindsey a gold lamé jacket, Frankie an animal-print coat. When everyone was dressed to their satisfaction, they headed uptown, where they promptly began drinking and dancing in her memory.

Tioga responded in kind, buying them rounds of shots. It was the best family reunion Robin McIntyre had ever attended; as she danced her way across the floor, one of the oil-boom newcomers stopped her.

"That's some celebration!" he said. "What's the occasion?"

"My grandma died!" she said, and kept dancing.

Back at the house, Savannah played with her cousin Claire, Lori's granddaughter.

"Where did you sleep last night?" Claire asked Savannah.

"In Grandma Gail's room."

"But Grandma Gail died," Claire protested.

"Yeah, her heart was broken," Savannah agreed.

"Maybe it came unplugged," Claire offered.

When a friend from church returned Gail's purse, all it contained was a wallet and a church bulletin. Her other purse remained on her nightstand;

behind it were the nitroglycerin pills that Chelsey Determan was always nudging her to remember.

A week later, when Karla was driving home to Fargo, she stopped near Jamestown, where her father had gone that terrible summer of 1978. She began sobbing as she had never done before, and she called her sister-in-law, Deb.

"I just want to call Mom right now," she wept. "I just want to talk to her, and I can't."

In the months since Gail's death, the ups and downs of her extended family's life continued. In North Dakota, as McKenna DeMoe neared the end of her senior year of high school and Lindsey, Dean's older daughter, put the finishing touches on her wedding, Deb was still commuting to Tioga on weekends. Dean's brain scans showed no changes that year, and he went out to celebrate with Tyler; but privately, some of his family members were worried. He was working too hard, as he always had. His temper was flaring sometimes, too—like when Kassie and Chelsey Determan took some of Gail's Christmas decorations to hang around Doug's and Lori's rooms in the nursing home. Dean came home from work after a particularly hard day and saw them in the driveway, laughing and chattering as they hauled out the boxes.

"Who said you could take this stuff?" he demanded. In fact, it had been Dean himself who'd said they could. He just didn't remember.

The girls were surprised, not just by the fact that Dean had forgotten, but by the idea that he even minded. He was constantly worried now that people were taking things from Gail's house—tchotchkes that he'd always derided as junk.

"Who cares?" Karla told him. "We *want* them to take it."

But now that Gail was gone, those little trinkets—her bits of junk, her sentimental sayings, even her mountains of holiday decorations—were like little pieces of her being carted away. Dean wanted to hold on to her memory for as long as he possibly could.

In May 2014, at Karla's urging and with Deb's support, Dean began participating in the anti-amyloid drug trial for the DIAN study in St. Louis.

The following year, on May 27, 2015, he became the first person in the study to be scanned for the presence of tau tangles, though the study did not disclose his results to him.

After the tau scan, he and Deb planned to visit St. Louis's Gateway Arch. But a nurse warned him that he was shedding so much radiation that he might get kicked out by security. Dean's eyes lit up.

"Let's go!" he told Deb. It would be like old times, when he got thrown out of places for breaking the rules.

By early 2016, the DIAN trial had scanned fifty brains for tau using T807—the same tracer that Reisa Sperling and Paul Aisen were using in their A4 study for late-onset Alzheimer's. The trial hoped to expand to additional countries, and Bateman said he hoped eventually to add anti-tau drugs to the clinical trials, though their development was a few years behind that of anti-amyloid medications.

"The current thinking is you need the cortex loaded with amyloid to get to the point where you unleash the tau-and-tangle cascade," said Pierre Tariot, who designed the Banner Institute's drug trials in the Colombian *paisa*. "It's the two, in series, that lead to the downstream effects, including brain cell dysfunction and death."

The Banner trial in Colombia was also scanning for tau, though unlike DIAN, it did not plan to introduce anti-tau drugs but rather stay only with crenezumab, which targets amyloid, throughout the duration of the study.

"The hope is by intervening with amyloid-clearing agents before the tangle cluster gets unleashed, [we] will delay or prevent it altogether," Tariot explained.

Ten months after Gail's death, a research team reported a discovery that could prove to be the long-sought link between mutation carriers and late-onset Alzheimer's patients. Led by Howard Federoff, executive dean of Georgetown University's School of Medicine, the team announced that it had developed a blood test that checked for unusually low levels of ten specific fats, and could predict, with 90 percent accuracy, whether a seemingly healthy person would develop Alzheimer's.

Because the test identified the disease risk before people developed symptoms, scientists believed they now had a way to know who might benefit from an early-intervention drug, if one succeeded in any of the trials that were under way.

While acknowledging that his test predated a proven preventative treatment for the disease, Federoff said he himself would still want to know if he was going to get it, so he could tie up loose ends in his life. He knew others might feel differently.

The issue of whether and how to tell people their odds of developing Alzheimer's was becoming more pressing as science inched closer to a viable treatment. In December 2015, the Banner Institute launched GeneMatch, a program that offered genetic testing for ApoE4 to people from the general public between fifty-five and seventy-five years old. After submitting a DNA sample from a cheek swab, participants would be matched to study opportunities. Some of those studies would allow people to learn their status through genetic counseling and would then follow them for a year to observe how the news impacted people psychologically.

Tariot and Reiman said the GeneMatch results would help providers determine the best way to disclose such sensitive news, particularly since ApoE4 testing is expected to become much more widespread if a treatment is found.

In March 2015, after much badgering by Karla, Jamie DeMoe underwent a week of scans and cognitive testing in Pittsburgh before receiving his first shot as part of the DIAN clinical trial. In keeping with study protocol, he would not know whether he received an active drug or a placebo.

Jamie had been through a trying year. The oil boom was over, and with prices falling, he had endured several layoffs in the oil field. He was now working in a plant that provided water for fracking. It was a boring job, but at least he was inside, and he got a respite from the hard physical labor of earlier years.

He arrived in Pittsburgh worried and frazzled by a lack of sleep, but as the week progressed, he grew more relaxed. On his last night, hours after receiving his first injection, he went out to dinner with Chelsey. They sat at the bar waiting for their table, and when they were called, he forgot his jacket; Chelsey gently reminded him.

On their way home to the hotel that night, he stopped in his tracks.

"I think I left my jacket back at the restaurant," he said. Chelsey realized he was right, and they returned for it. He smiled broadly. *He* had remembered.

That same week, at a conference in Nice, France, pharmaceutical company Biogen announced that it was expanding its human clinical trials of the anti-amyloid drug aducanumab after a small study had yielded promising results in older patients.

The trial enrolled 166 people, most in their early seventies, whose diagnosis was confirmed when their brain scans—using a derivative of Bill Klunk's and Chet Mathis's PiB—revealed amyloid buildup. Clinically, about 60 percent showed mild symptoms, and two-thirds of them carried at least one ApoE4 allele.

After a year, the control group's amyloid buildup rose slightly, while people in all groups that received the drug experienced a drop, the sharpest of which occurred in the people taking the highest dose. In cognitive testing, the control group's score fell three points after a year, while the group in the highest drug dosage dropped by less than one point and seemed to stabilize.

But there were side effects, too, especially at the highest doses: specifically, vasogenic edema, or ARIA, the same fluid on the brain which had plagued other anti-amyloid drugs. In a few of the patients, this caused headaches, confusion, and visual disturbances, but those symptoms usually resolved within a month, according to the Biogen representative who presented the findings.

Klunk and Tariot both expressed cautious enthusiasm for the findings, but were careful to add that the study was very small and that previous drugs had failed in the past when they moved to larger clinical trials. However, the results also offered hope that they were on the right track, and that clearing amyloid would help stave off cognitive decline.

"A win for one agent is likely a win for the others that are viable," Tariot said.

Paul Aisen, whose A4 study was about halfway to its goal of enrolling a thousand people, was also pleased.

"It's preliminary, but I think very exciting," he said. "That's made me think we want to get anti-amyloid therapy to people as early as possible, but it looks like it may be helpful even at the symptomatic stage."

Days later, the Mayo Clinic published findings that reminded the field that tau was still very much in play. The NIH-funded study, which looked at thirty-six hundred brains postmortem, identified tau as the primary driver behind memory loss in Alzheimer's disease, though it did not rule out the possibility that amyloid played a role.

Toward that end, some pharmaceutical companies—including Biogen and AC Immune, the company that created crenezumab, the anti-amyloid drug Banner was using in Colombia—continued developing drugs to target tau, opening the door to the possibility of eventually pursuing combination therapies that treated both plaques and tangles, the "holy grail" that Reisa Sperling had described.

Aisen, though still a Baptist, was persuaded that tau was a worthy target. In 2015, the A4 study—like DIAN—began scanning patients' brains for tau. Combined with the amyloid scans, it showed that tau levels continued to rise later in the disease, after amyloid had leveled off, and seemed to correlate more closely with the severity of the person's symptoms.

But Aisen said the scans also showed that tau levels, like amyloid, were rising before symptoms started, suggesting that when researchers did start offering anti-tau drugs, they should start early, and that anti-tau medication may be helpful throughout the disease, not just at the beginning.

"I believe that the driver is amyloid, but that doesn't mean amyloid is the best therapeutic target," Aisen said. "I've placed my bet on amyloid. But until we're successful, we have to pursue every plausible strategy."

Reiman recalled his dismay when he first heard the government set 2025 as the target date for finding a viable Alzheimer's treatment. He feared that the field was risking its credibility: What if they didn't meet that goal?

But given the progress that had been made just in the past few years, he had changed his mind. Now he was gaining confidence that the trials were working. So many people had sacrificed so much to get there. Doctors had risked their careers. People like the DeMoes had bet their lives on it. When nobody cared about this disease, they had refused to fade away.

"There is a strategic path forward," Reiman said. "The stars are going to need to align, but we have a chance."

As she learned about each new development in the field, Karla thought about how thrilled her mother would have been to see the progress that had been made in just the few years since her death.

"To think, when she started, there was just nothing," Karla said. "She would be so, so proud of the kids and the grandkids that are participating."

Deep in her heart, she hoped Gail was proud of her, too, and of her efforts to save her siblings, her nieces and nephews, her cousins. Once, she was the girl who refused even to go to the grocery store for her mother; now she had taken up Gail's mantle. She wondered what her father would think of her.

"Growing up with somebody with the disease, I was really dependent on other people. I needed my friends, needed Matt. And now I feel I can do this thing," Karla said. "I guess since I've done it without Mom now, I feel like I can do it. We can go the distance. There's so much progress, there is so much hope."

In the photographs, the notes, the home movies, and the memories she left behind, Gail loomed large in the lives of all she touched, emboldening them to keep fighting. To a person, her family, in the moments when they paused long enough to feel her presence, imagined they could hear her signature line:

I love you to the moon and back.

Acknowledgments

To allow a stranger complete access to the most intimate details of your family history requires a leap of faith that, in all honesty, I'm not certain I would have the courage to make. Yet that is exactly what the DeMoe family did for me, and without their trust, their story would have been impossible to tell. They showed me every bump and wrinkle, every triumph and defeat, and they welcomed me as their own. I can never repay them for that privilege. I can only hope that the end result, which not all of them survived to see, is worthy of that trust. You are *paisa*.

To my agent, Larry Weissman, and his wife and business partner, Sascha Alper, thank you for believing in this story and championing it every step of the way. Thanks also to author Dave Kinney, my onetime AP colleague, for introducing me to Larry and sharing with me the process for pitching a book.

Millicent Bennett was my first editor at Simon & Schuster, and I am indebted to her for shepherding this manuscript, pushing me beyond what I thought were my limits, and for caring about its characters as much as I do. Karyn Marcus, who became my second editor, helped me see the story through a new lens and bring it to fruition, culminating years of planning and research. Thank you, both.

To the brilliant minds whose lives have been spent in pursuit of a single goal—to rid the world of Alzheimer's—I am humbled by, and grateful for, your generosity in helping me tell this story. Bill Klunk, Chet Mathis, Snezana Ikonomovic, and the entire staff of the University of Pittsburgh's Alzheimer's Disease Research Center—thank you for the many hours we spent together. From the DIAN team, I would like to thank Eric McDade (first of Pitt, now of Washington University), Randy Bateman, and John Morris. From the Banner team, I thank Eric Reiman,

Jessica Langbaum, and Pierre Tariot; their accessibility, even in the midst of this most important work, was both impressive and critical. Being able to relate such complex scientific information in an elegant way is a rare gift, and you have it. To Teresa Zyznewsky, I am grateful for your translation help, which allowed me to reach out to Dr. Lopera; to Francisco Lopera and Ken Kosik, I thank you for sharing your adventures with me, and I hope to one day join you. I thank Ken also for the many times you clarified and expanded upon your work and for your seemingly limitless patience. I thank Paul Aisen and Reisa Sperling of the A4 trial: With minds such as yours working on this problem, I can't help but be filled with hope. I am grateful to June White, Peter St. George-Hyslop, and Jerry Schellenberg for their willingness to speak to me, and to Dmitry Goldgaber for sharing his remarkable story.

I am indebted to Becky Vork and Monte Olsen, who helped me understand what true friendship means and restored my faith in its existence. Monte, thank you for the dance. I am grateful to Julie Noonan Lawson and Kate Preskenis for speaking to me about their family's experiences.

Writing can be a lonely occupation, but for the shared experiences of those who have been there before. Thank you to Fawn Fitter, who included me in her acknowledgments and to whom I am finally returning the favor; and to Cindi Lash and Chuck Finder, for their counsel and friendship. The best writers I ever knew were the journalists of the Associated Press in Concord, Los Angeles, and Pittsburgh. There are too many of you to name individually, but I thank you all the same for showing me the way.

I am blessed with resourceful friends who were kind enough to help me with my research, particularly Stuart Culy, who tracked down congressional testimony, and Tina Wood, without whom I might never have found the elusive June White. Julian Neiser's legal advice and assistance in locating documents related to Trey Sunderland were immensely helpful, and he's also a talented journalist when he's not busy practicing law.

My friend Sam Menchyk was essential in uploading and, in many cases, restoring the photographs that illustrate this book. He has also been indispensable in advising me about the digital storytelling process as I develop the book's website. I can never adequately thank him for his expertise and generosity.

I thank my own family and friends: my father, George Kapsambelis, who made me believe I could do anything because he believed it; my mother, Leslie Kapsambelis, a writer who wanted me to follow in her footsteps and celebrated each step of that journey; and my grandmother and personal hero, Helen Kapsambelis, who taught me the lengths to which a parent will go to save her children. To my nuclear family: Dave Adler, who was integral to the logistics of this project and took care of my most precious assets, Bobby and Tim Adler, while I traveled.

To Bobby and Tim: Simply stated, you are my reason for being. Nobody will ever be more important to me than you are.

Last, I thank Craig Brickell, who always told me the truth and sat across the table from me one evening at a bar in Pittsburgh and told me I had to write this story.

A Timeline of Discovery in Alzheimer's Disease

1900s

1906: Alois Alzheimer, a German psychiatrist, identifies a new disease in a fifty-one-year-old woman named Auguste Deter. Alzheimer's boss later names the disease after him.

1940s

1948: A London doctor named R. D. Newton reviews the previous forty years' worth of literature and concludes that Alzheimer's and old-age senility are the same disease. He also guesses that a hereditary pattern is at play.

1960s

1963: Neurologist Robert G. Feldman, first at Yale and later at Boston University, writes a paper about a family whose history with Alzheimer's seems to suggest a pattern of genetic inheritance.

1968: British scientists Gary Blessed, Bernard Tomlinson, and Martin Roth write a seminal paper concluding, once and for all, that dementia in older people is the same disease as Alzheimer's in younger people. This is one of the first hints indicating how widespread the Alzheimer's problem actually is.

1970s

1972: Jean-François Foncin, a French neuropathologist, diagnoses a young mother with Alzheimer's. By tracing her ancestry, he discovers a large extended family who carry a mutation that causes Alzheimer's disease. He later realizes they are the same family who formed the basis for Robert Feldman's research. They come to be known in medical literature as "Family N."

1976: Robert Katzman, a neurologist and medical activist from the University of California San Diego, identifies Alzheimer's disease as a "major killer" in an editorial he writes for the American Medical Association's *Archives of Neurology*.

1980s

1980: Katzman helps found the Alzheimer's Association. From 1980 to 1996, federal funding for Alzheimer's research increases from $5 million to $300 million.

1984: George Glenner, a pathologist from the University of California at San Diego, identifies the exact makeup of the protein that forms the amyloid plaques found in Alzheimer's patients. Known as amyloid beta, they are fragments of a protein manufactured by chromosome 21, the same chromosome that is duplicated in people with Down syndrome.

1984: Colombian neurologist Francisco Lopera meets his first patient suffering from genetically inherited Alzheimer's disease. The patient turns out to be a member of the largest-known family tree of people with this form of the disease.

1985: Peter St. George-Hyslop, a British neurologist and molecular geneticist, partners with Jean-François Foncin to isolate the genetic origins of Alzheimer's disease in Family N.

1986: Dmitry Goldgaber, a Russian-born scientist working at the NIH, discovers the first-known mutation that causes Alzheimer's. Known as the APP gene, it is located on chromosome 21, right where George Glenner predicted it would be. But it accounts for only a small fraction of the people who are inheriting Alzheimer's disease through a genetic mutation.

1989: Leonard Heston at the University of Minnesota contributes to a brain autopsy of Galen "Moe" DeMoe. His collaborator, a nurse named June White, collects DNA from several DeMoe family members for Heston's research exploring the possibility of a mutation different from the APP mutation located by Dmitry Goldgaber.

1990s

1992: Heston and White collaborate with biochemist Jerry Schellenberg at the University of Washington. Schellenberg narrows the search for a new mutation down to a segment of chromosome 14. The results are published in *Science* magazine.

1992: Lopera teams up with Harvard neuroscientist Ken Kosik to pursue research with the extended Colombian family and seek a cure.

1992: Allen Roses at Duke University identifies variations, which are called alleles, of the ApoE gene on chromosome 19 that heighten the risk of Alzheimer's disease within the general population. Roses becomes convinced that the tau protein, not amyloid beta, is the chief culprit behind Alzheimer's disease.

1995: St. George-Hyslop discovers a second mutation, which he names presenilin 1 (PS1), on the section of chromosome 14 previously identified by Schellenberg. PS1 turns out to be the most common genetic mutation in Alzheimer's disease, afflicting the DeMoes and Family N, as well as Alois Alzheimer's original patient, Auguste Deter.

1995: Schellenberg and others, including Rudolph Tanzi at Harvard, discover a third mutation—presenilin 2 (PS2)—on chromosome 1.

2000s

2002: After years of failed experiments, Bill Klunk, a psychiatrist, and radiochemist Chet Mathis—both at the University of Pittsburgh—develop a chemical known as "Pittsburgh Compound B," or PiB, which allows science to view the progression of Alzheimer's disease in a living brain for the first time, a critical tool in testing drugs that fight amyloid beta.

2006: Eric Reiman, a brain imaging expert and neuroscientist, partners with Pierre Tariot, a psychiatrist specializing in the design of clinical trials, to create an ambitious program through the Banner Alzheimer's Institute in Phoenix to detect and attack Alzheimer's in its earliest stages, before a person develops symptoms.

2006: Pearson "Trey" Sunderland III, chief of geriatric psychiatry at the National Institute of Mental Health, pleads guilty to criminal conflict of interest for his relationship with pharmaceutical giant Pfizer. His research, which includes samples drawn from members of the DeMoe family, is put on hold. The DeMoes transfer their efforts to the University of Pittsburgh, where they meet Klunk, who begins imaging their brains.

2008: John Morris at Washington University launches an international study of people who carry the known mutations causing Alzheimer's disease. Through Klunk, the DeMoes enroll in Morris's study, known as the Dominantly Inherited Alzheimer's Network (DIAN). The study names neurologist Randy Bateman to lead its planned drug-trial unit.

2010: Reiman and Tariot establish a formal collaboration with Lopera and Kosik to test an anti-amyloid prevention drug for Alzheimer's in the Colombian families identified by Lopera.

2012: The Obama administration sets a goal of finding a way to prevent and treat Alzheimer's disease by 2025.

2012: DIAN's trial unit begins testing two anti-amyloid drugs in people with an inherited genetic mutation, including members of the DeMoe family.

2014: Harvard neurologist Reisa Sperling and her longtime collaborator, Paul Aisen at the University of Southern California, launch a study testing one of the DIAN trial's anti-amyloid drugs in people from the general population who are sixty-five to eighty-five years old. The study, known as A4, will also scan participants for the tau protein in hopes of one day adding an anti-tau drug to the trial.

2015: The DIAN trial begins scanning participants' brains for the presence of the tau protein, which some researchers believe is more critical to the progression of Alzheimer's symptoms than amyloid. Dean DeMoe is the first person scanned.

2015: The Alzheimer's Association awards $10 million in new research funding to a trial launched by Reiman, Tariot, and the Banner Alzheimer's Institute's Jessica Langbaum testing two additional drugs in people ages sixty to seventy-five who carry two copies of the ApoE4 gene variant discovered by Roses in 1992. The study also plans to image participants' brains for the tau protein, as does the study focusing on the Colombian family.

2015: Reiman, Tariot, Langbaum, Lopera, Bateman, Morris, Sperling, Aisen, and Roses are coauthors on a paper describing the Collaboration for Alzheimer's Prevention, an initiative that allows them to share information gathered in their respective trials.

2025: The US target date for discovering a prevention and treatment of Alzheimer's disease.

2050: Projected number of Americans with Alzheimer's if a cure is not found: 13.8 million. Projected number worldwide: 131.5 million.

Notes

ONE: THE ENEMY WITHIN

3 In the United States, Alzheimer's is the sixth-leading cause of death: "2016 Alzheimer's Disease Facts and Figures," Alzheimer's Association.

3 Next to cancer, there is no condition more feared: Harris poll for MetLife Foundation, September 2010.

3 An estimated 24 to 36 million people worldwide—5.3 million in the United States alone: Martin Prince et al., "World Alzheimer's Report 2015: The Global Impact of Dementia," Alzheimer's Disease International, reports that more than 46 million people worldwide have dementia. The report "2016 Alzheimer's Disease Facts and Figures," produced by the Alzheimer's Association, estimates that 60 to 80 percent of all dementia cases are caused by Alzheimer's disease.

4 Only one in four people who have the disease are actually diagnosed: "The World Alzheimer Report 2011," Alzheimer's Disease International.

4 Once thought to be relatively rare: "Understanding Alzheimer's," United Healthcare; "2016 Alzheimer's Disease Facts and Figures." Alzheimer's Association.

4 Alzheimer's is now known to be the leading cause of age-related dementia: Ibid.

4 The disease was first identified in 1906 by its namesake, Alois Alzheimer: Sources used for the story of Alois Alzheimer and his patient, Auguste Deter, include: Konrad Maurer and Ulrike Maurer, *Alzheimer: The Life of a Physician & The Carrier of a Disease,* translated by Neil Levi with Alistair Burns (New York: Columbia University Press, 1998); David Shenk, *The Forgetting: Alzheimer's: Portrait of an Epidemic* (New York: Doubleday, 2001); Peter J.

Whitehouse with Daniel George, *The Myth of Alzheimer's: What You Aren't Being Told About Today's Most Dreaded Diagnosis* (New York: St. Martin's Press, 2008); Márcio Borges, "A História de Auguste Deter," Cuidar de Idosis, June 24, 2011; Hanns Hippius and Gabriele Neundörfer, "The Discovery of Alzheimer's Disease," *Dialogues in Clinical Neuroscience*, March 2003.

4 In the second century, Roman emperor Marcus Aurelius: "Greek Medicine," History of Medicine Division, National Library of Medicine, National Institutes of Health, first published September 16, 2002.

5 Doctors relied on clinical tests: "2016 Alzheimer's Disease Facts and Figures," Alzheimer's Association, p. 14.

5 A study of 852 men diagnosed with Alzheimer's disease: Robin Erb, "Diagnosis of Alzheimer's Isn't Always Accurate," *Detroit Free Press*, May 17, 2012; "Experts Are Concerned About Misdiagnosis Rate for Alzheimer's Disease," *McKnight's*, May 18, 2012.

6 Only about 1 percent of all Alzheimer's patients: "2016 Alzheimer's Disease Facts and Figures,"Alzheimer's Association, p. 10.

6 their average age of onset is between thirty and fifty years old: "Younger/Early Onset Alzheimer's & Dementia," Alzheimer's Association.

7 had invented a method for staining brain cells: David Shenk, "The Memory Hole," *New York Times*, November 3, 2006.

9 "I have lost myself": Ibid.

9 He attributed their dementia to atherosclerosis: Maurer and Maurer, *Alzheimer: The Life of a Physician*, p. 17.

11 Ironically, Kraepelin—who valued classification—unwittingly worsened a key confusion: Jesse F. Ballenger, "Progress in the History of Alzheimer's Disease: The Importance of Context," *Alzheimer's Disease: A Century of Scientific and Clinical Research*, edited by George Perry, Jesus Avila, June Kinoshita, and Mark A. Smith (Amsterdam: IOS Press, 2006), p. 7; José Manuel Martinez Lage, "100 Years of Alzheimer's Disease (1906–2006)," *Alzheimer's Disease: A Century of Scientific and Clinical Research*, pp. 19–20.

12 Alzheimer himself didn't dispute this: Ballenger, "Progress in the History," p. 7.

12 were added to blue cardboard files and left to collect dust deep within Johann Wolfgang Goethe Frankfurt University Hospital: Whitehouse, *The Myth of Alzheimer's*, p. 78.

TWO: THE SALT OF THE EARTH

14 Excerpts of letters from Galen DeMoe used with the permission of Gail DeMoe.

15 "It's funny how at that age, sparks can fly": Author's interview with Gail DeMoe, August 24, 2011.

15 The self-proclaimed "oil capital of North Dakota," Tioga lies nestled in the state's northwest corner: "History of Tioga," www.tiogand.net.

16 the Amerada Petroleum Corporation discovered oil in Tioga: Details about the Bakken Formation and the history of the oil industry in the region were drawn from "The Bakken Boom: An Introduction to North Dakota's Shale Oil," Energy Policy Research Foundation, Inc., August 3, 2011; "Son of Bakken Formation Namesake Remains Reserved," Associated Press, December 3, 2012; and James Vlahos, "Oil Boom: North Dakota is the Next Hub of U.S. Energy," *Popular Mechanics*, June 13, 2012.

18 "As far as Moe goes": Author's interview with Hank Lautenschlager, July 2012.

19 He and his best friend, Gary Anderson: Author's interview with Gary Anderson, October 20, 2015.

20 "She drove this old car": Author's interview with Karla Hornstein, August 17, 2011.

21 One officer, who also happened to be Doug's boxing coach: Author's interview with Gary Anderson.

21 Gail herself had seen other changes: Author's interview with Gail DeMoe.

22 To avoid parallel parking: Details of Wanda DeMoe's illness were drawn from the author's interviews with Gail DeMoe, February 14, 2013; Victor DeMoe, September 13, 2012; and Rob Miller, March 31, 2014.

22 Whatever was ailing Wanda also seemed to be affecting her two brothers: Author's interview with Gail DeMoe, February 14, 2013.

22 When Wanda's husband took her to a doctor, she was prescribed shock treatments: Ibid.

22 Bedsores covered her hips, buttocks, and shoulders: Autopsy report of Ruth (Wanda) DeMoe, Sacred Heart Hospital, Eau Claire, Wisconsin; autopsy no. 44668, July 15, 1964.

23 And in 1963, the year before she died, a group of doctors led by Robert Feldman: R. G. Feldman et al., "Familial Alzheimer's Disease," *Neurology* 13, no. 10 (1963).

23 Jerry adapted quickly to his oldest brother's household: Author's interview with Karla Hornstein, February 11, 2015.

23 Kindhearted neighbors took in Moe's youngest brother, Ray: Ibid.

23 On March 25, 1967, Ray's platoon was pinned down by enemy fire: "Elk Mound Marine Killed in Vietnam," *Eau Claire Leader*, March 30, 1967.

25 At work, Hank Lautenschlager was slow to notice any changes: Author's interview with Hank Lautenschlager.

25 Eventually, in 1974, Moe was let go from the job: Comprehensive History and Record Review of Galen DeMoe by the St. Cloud VA Medical Center, St. Cloud, Minnesota, July 16, 1981.

25 Karla, ever conscious of appearances, was so ashamed: Author's interview with Karla Hornstein, February 11, 2015.

26 In the early Sunday morning hours of a summer heat wave: Letter from Lori DeMoe to Steve McIntyre.

28 "You're just trying to make it through it all yourself": Author's interview with Karla Hornstein, February 11, 2015.

28 In August 1978: Details of Moe's last night at home were drawn from the author's interviews with Karla Hornstein, Dean DeMoe, Gail DeMoe, and Hank Lautenschlager.

THREE: FAMILY N

30 Things began to change in 1948: Daniel A. Pollen, *Hannah's Heirs: The Quest for the Genetic Origins of Alzheimer's Disease* (New York: Oxford University Press, 1993), pp. 55–59.

30 including the one written by Harvard's Robert Feldman in 1963: Pollen, *Hannah's Heirs*, pp. 143–144; Feldman et al., "Familial Alzheimer's Disease."

31 But the turning point came in 1968: Gary Blessed, Bernard Tomlinson, and Martin Roth, "The association between quantitative measures of dementia and of senile change in the cerebral grey matter of elderly subjects," *British Journal of Psychiatry*, July 1968; Kenneth S. Kosik and Ellen Clegg, *The Alzheimer's Solution: How Today's Care Is Failing Millions and How We Can Do Better* (New York: Prometheus Books, 2010), p. 88.

31 It was into this beautiful, sometimes brutal land that French neuropathologist Jean-François Foncin landed in May 1973: Jean-François Foncin, "A Neuropathologist and Alzheimer Genetics," *Alzheimer: 100 Years and Beyond,* edited by George Perry, Jesus Avila, June Kinoshita, and Mark A. Smith (New York: Springer, 2006), pp. 225–230.

34 such was reportedly the case with Feldman: Rudolph E. Tanzi and Ann B. Parson, *Decoding Darkness: The Search for the Genetic Causes of Alzheimer's Disease* (Cambridge: Perseus, 2000), pp. 58–59.

34 Feldman did the best he could to analyze the family's heredity with mid-1960s technology, but it wasn't enough: Pollen, *Hannah's Heirs*, p. 144.

34 Their 1985 paper appeared to offer: Ibid.

35 Hyslop found write-ups of a Canadian family: Author's interview with Peter St. George-Hyslop, April 4, 2013; Pollen, *Hannah's Heirs*, pp. 142–143,

36 The same committee members who had cut his grant nine years earlier: Foncin, "A Neuropathologist and Alzheimer Genetics," p. 228.

FOUR: ONE IN A MILLION

37 Dr. Robert Katzman, a neurologist and medical activist from the University of San Diego: José Manuel Martínez Lage, "100 Years of Alzheimer's disease (1906–2006), *Journal of Alzheimer's Disease* 9 (2006) 15–26; H. Roger Segelken, "Robert Katzman, Alzheimer's Activist, Dies at 82," *New York Times*, September 23, 2008.

38 From 1980 to 1996, federal funding for Alzheimer's research increased: Debra Kain, "Dr. Robert Katzman, Pioneering Alzheimer's Disease Expert, Dies," press release from the UC San Diego News Center, September 18, 2008.

38 Dmitry Goldgaber was up for the challenge: Biographical details about Dmitry Goldgaber drawn from the author's interview with Dmitry Goldgaber, February 21, 2013; Dmitry Goldgaber, "My Story: The Discovery and Mapping of Chromosome 21 of the Alzheimer Amyloid Gene," *Journal of Alzheimer's Disease* 9 (2006) 349–360. Also appears in *Alzheimer's Disease: A Century of Scientific and Clinical Research*, pp. 349–360 (see chap. 1, n. 16).

40 Gajdusek had built a career: Donald G. McNeil Jr., "D. Carleton Gajdusek, Who Won Nobel for Work on Brain Disease, Is Dead at 85," *New York Times*, December 15, 2008.

40 Goldgaber was ecstatic: Author's interview with Dmitry Goldgaber.

41 Carleton Gajdusek wondered if Alzheimer's: Goldgaber, "My Story."

41 Prions are protein particles: Descriptions of prion disease aided by the "prion disease" entry in the *Genetics Home Reference* (National Library of Medicine, January 2014); Jim Schnabel, "Alzheimer's Disease: Return of the Prion Hypothesis," The Dana Foundation, February 17, 2012.

42 In 1980, Gajdusek tested the theory: McNeil, "D. Carleton Gajdusek."

42 In 1984, a pathologist named George Glenner: Wolfgang Saxon, "Dr. George G. Glenner, 67, Dies; Researched Alzheimer's Disease," *New York Times*, July 14, 1995; Kosik and Clegg, *The Alzheimer's Solution*, pp. 90–92.

42 He found them to be much smaller: Schnabel, "Alzheimer's Disease: Return of the Prion Hypothesis."

42 Virtually all people with Down syndrome: "Down Syndrome and Alzheimer's Disease," Alzheimer's Association, 2015.

43 But back in the early 1980s, there was no database to search: Kosik and Clegg, *The Alzheimer's Solution*, p. 91.

43 Gajdusek liked to excite them with scientific conundrums: Author's interview with Dmitry Goldgaber.

43 Had he known, Goldgaber never would have bothered: Ibid.; Goldgaber, "My Story."

45 When the amyloid precursor protein (APP) gene is mutated: *Genetics Home Reference.*

46 Goldgaber showed up and, after some negotiation, was allowed to pre-empt the program: Goldgaber, "My Story"; Tanzi and Parson, *Decoding Darkness*, pp. 79–81; Kosik and Clegg, *The Alzheimer's Solution*, p. 92.

FIVE: YOU ARE MY SUNSHINE

48 "I remember asking my mother": Author's interview with Gail DeMoe, July 2012.

49 she suffered the first of a series of nervous breakdowns: Author's interview with Karla Hornstein, May 6, 2016.

49 which Gail euphemistically referred to as "a learning experience": Author's interview with Gail DeMoe, March 5, 2013.

50 St. Cloud would become Moe's home for the next seven years: Medical record of Galen DeMoe, St. Cloud VA Medical Center, December 5, 1978–January 8, 1986.

50 "I want to go home and be with my kids": Medical record notes, May 7, 1979.

50 A nurse reported that he was "sitting in the dayroom sobbing as though his heart would break": Medical record notes, August 1979.

50 By August 1980, Moe could not remember the names of all his offspring: Medical record notes, August 1980.

50 By April 1985, he could not consistently respond to his own name: Medical record notes, April 3, 1985.

50 He sat strapped into a chair designed for geriatric patients: Ibid.

50 "I think the day Dad left": Author's interview with Karla Hornstein, January 27, 2013.

51 *I don't want you and Mother to figure that it was your fault*: Undated letter from Brian DeMoe to his father, postmarked February 19, 1975.

51 If he'd asked Debbie to marry him back then, she would have: Author's interview with Debbie (Thompson) Ness, April 10, 2014.

52 They met at a dance in the eighth grade: Author's interviews with Karla and Matt Hornstein, July 2012.

53 "I don't know what draws people together": Author's interview with Matt Hornstein, October 8, 2015.

55 He verbally berated Christy: Author's interview with Christy Holm, July 2012.

55 But there was also a side to Brian that endeared him to people, especially Gail: Author's interview with Gail DeMoe, July 2012.

56 Hydergine, an extract of a fungus that grows on rye: "Summary of Data for Chemical Selection: Hydergine," Technical Resources International, Inc., December 1999.

56 The drug was developed by Albert Hofmann: Ibid.

56 "I understand that the reports and data collected from this investigation may not directly benefit me in the treatment of my disorder": Information About the VA Cooperative Study (#32) on Pharmacotherapy of Chronic Organic Brain Syndrome, VA Form 10-3518-22, January 1978; signed by Galen DeMoe on April 5, 1979.

57 The study enrolled patients from nine VA facilities: Details of the study in which Moe was enrolled are from Grant Huang, acting director, Cooperative Studies Program, Department of Veterans Affairs Office and Research Development, in an email to the author dated March 27, 2015.

57 with most concluding that its benefits were extremely modest at best: T. L. Thompson et al., "Lack of Efficacy of Hydergine in Patients with Alzheimer's Disease," *New England Journal of Medicine*, August 16, 1990; J. Olin et al., "Hydergine for Dementia," *Cochrane Database of Systematic Reviews*, 2001.

SIX: THE GHOSTS OF ANOKA

59 Though Karla knew her grandmother also had the disease: Author's interview with Karla Hornstein.

59 Karla was flattered by the researchers' attention: Author's interview with Karla Hornstein.

59 A mother of four, June White began her career as a registered nurse: Author's interview with June White, April 12, 2013.

60 Heston, a psychiatrist and geneticist: For more background on Leonard Heston, see Robert Katzman and Katherine Bick, *Alzheimer Disease: The Changing View* (Cambridge, MA: Academic Press, 2000); entry on Heston, including an interview by Robert Katzman, beginning on p. 150.

60 Before she left, she called Heston: Author's interview with June White.

62 he was moved to the Americana Nursing Home: Medical record of Galen DeMoe, St. Cloud VA Medical Center, January 8, 1986.

62 Karla hated to admit it: Author's interview with Karla Hornstein.

64 The autopsy report was dated June 9, 1989: Autopsy report of Galen DeMoe.
65 "It was an exciting project": Author's interview with Peter St. George-Hyslop, April 4, 2013.
66 occurring in only 1 to 5 percent of all Alzheimer's patients in the world: National Institute on Aging, "Alzheimer's Disease Genetics Fact Sheet," accessed February 2, 2015.
66 "We were literally plucking markers out of the air": Author's interview with Jerry Schellenberg, March 29, 2013.
67 Schellenberg was waiting for a printout: Author's interview with Schellenberg; Tanzi and Parson, *Decoding Darkness*, pp. 147–148.
67 Karla received a note: Letter from Leonard Heston and June White to Karla Hornstein, October 20, 1992.
68 The letter came with a copy of the *Science* article: Additional coverage of the discovery included Jean Marx, "Familial Alzheimer's Linked to Chromosome 14 Gene," *Science* 258, October 23, 1992.
69 Karla wasn't really sure what any of it meant: Author's interview with Karla Hornstein.
69 even Auguste Deter: Ulrich Müller et al., "A Presenilin 1 Mutation in the First Case of Alzheimer's Disease," *Lancet Neurology*, December 14, 2012.
69 Yet Hyslop thought their lack of financial resources might have had an unexpected benefit: Author's interview with Peter St. George-Hyslop.
70 Hyslop waited until he was absolutely certain they'd identified the gene: Ibid.

SEVEN: UNTAMED HEARTS

71 "Grandma had her own secrets": Author's interview with Karla Hornstein, January 27, 2013.
72 Everyone in the oil patch has a nickname: Author's interview with Dean DeMoe, July 2012.
72 becoming dependent on them: Author's interview with Karla Hornstein.
72 Even so, she always thought of them as close: Author's interview with Jennifer DeMoe, July 2012.
73 In the nearly four decades Steve McIntyre spent working on the railroad: Author's interviews with Steve and Lori McIntyre, May 2012.
74 Every birthday cake the girls ever had: Author's interview with Robin McIntyre, September 2, 2012.
75 Dean regularly did the jobs of seven men: Author's interview with Deb DeMoe, March 16, 2012.

75 He sometimes jokingly referred to weddings as "another funeral": Author's interview with Tyler DeMoe.

75 Monte was driving a work truck: Author's interview with Monte Olson, March 7, 2014.

76 Dean finally met his match: Author's interviews with Monte Olson and Deb DeMoe.

77 The loser had to make milk shakes: Author's interview with McKenna DeMoe, July 2012.

77 "I never really got to hang out with him": Author's interview with Jamie DeMoe, September 12, 2012.

78 he struggled with learning disabilities: Author's interviews with Gail DeMoe, July 2012, and Karla Hornstein, January 23, 2012.

78 "You're the mistake!" they'd tell him: Author's interview with Tyler DeMoe.

78 She remembered driving past her mother's house: Author's interview with Karla Hornstein, January 23, 2012.

78 Most people, including Jamie himself, described him as a classic worrywart: Author's interview with Rikki Rice, July 2012.

EIGHT: A BLAMELESS AND UPRIGHT MAN

84 Gail's warning stuck with Sharon: Author's interview with Sharon DeMoe, December 3, 2012.

84 The truth was, small lapses were starting to add up to a more ominous picture: Author's interview with Sharon DeMoe.

85 Sharon sat Sheryl, her younger daughter, down: Sheryl Grammer, "My Journey."

85 in Bethesda Jerry underwent two weeks' worth of tests: Inpatient medical record of Jerry A. DeMoe, December 1, 1995–December 15, 1995, National Institutes of Mental Health.

85 both obesity and cardiovascular disease have been associated with a greater risk for Alzheimer's: "Be Heart Smart," Alzheimer's Association, accessed May 9, 2016.

86 researchers have been exploring the role of cholesterol: Luigi Puglielli, Rudolph E. Tanzi, and Dora M. Kovacs, "Alzheimer's Disease: The Cholesterol Connection," *Nature Neuroscience*, April 2003.

86 "In some ways, it was kind of a relief to him to know": Author's interview with Sharon DeMoe.

87 Sold under the brand name Cognex: Tacrine entry, Drugs.com, accessed March 3, 2014.

87 In 1991, an FDA advisory committee rejected Cognex: Gina Kolata, "F.D.A. Advisers Reject a Drug for Alzheimer's," *New York Times*, March 16, 1991.

87 At least one panel member called it "a matter of conscience": Thomas H. Maugh II, "Alzheimer's Drug Backed for Approval," *Los Angeles Times*, March 19, 1993.

87 Summers, who took a sabbatical to promote the drug's approval to the FDA: Ibid.

87 But Jerry never saw the improvements: Sheryl Grammer, "My Journey."

87 Soon after the diagnosis, Sharon made herself a vow: Author's interview with Sharon DeMoe.

87 It was a scary transition for Sheryl: Author's interview with Sheryl Grammer, January 24, 2012.

88 "a pretty and sweet season of my life, gone": Sheryl Grammer, "My Journey."

88 to his wife, he talked about suicide: Author's interview with Sharon DeMoe, January 22, 2016.

89 "I mostly wanted help": Author's interview with Sharon DeMoe, December 3, 2012.

89 "We knew the pain": Author's interview with Sharon DeMoe, January 22, 2016.

89 Sheryl was consumed with shame: Sheryl Grammer, "My Journey."

90 "I know how that goes": Author's interview with Sharon DeMoe, January 22, 2016.

92 Sharon called Sheryl to say: Come home: Sheryl Grammer, "My Journey."

NINE: THE FRUITS OF PERSISTENCE

93 In the years that followed, time became the chief roadblock: Author's interview with Eric Reiman, October 18, 2012.

93 The National Institutes of Health conducted its Alzheimer's Disease Anti-inflammatory Prevention Trial: Breitner et al., "Extended Results of the Alzheimer's Disease Anti-Inflammatory Prevention Trial," *Alzheimer's & Dementia*, July 2011; NIH press release, "Use of Non-Steroidal Anti-Inflammatory Drugs Suspended in Large Alzheimer's Disease Prevention Trial," December 20, 2004.

94 Since Alzheimer's patients had plaques forming outside their brain cells: For an explanation of how plaques and tangles form, see "Alzheimer's Disease: Unraveling the Mystery," National Institute on Aging, September 2008 and updated January 22, 2015.

95 this was the challenge facing Chet Mathis: Author's interviews with Chet Mathis and Bill Klunk, November 7, 2011, and January 27, 2012.

96 As its name suggests, it's a bright red, toxic sodium salt: David P. Steensma, "'Congo' Red: Out of Africa?" *Archives of Pathology*, February 2001, pp. 250–252.

97 One day, however, a young psychiatry professor named Bill Klunk: Author's interviews with Chet Mathis and Bill Klunk.

98 Kupfer had the faith to stand by him: Author's interview with David Kupfer, July 17, 2013.

99 The difference between success and failure sometimes rests on a single atom: *The Forgetting: A Portrait of Alzheimer's*, PBS, 2010, 66:22.

99 Mathis's ideal dye couldn't just leak into the brain in trace amounts: Author's interviews with Chet Mathis and Bill Klunk.

99 Klunk and Mathis started with Thioflavin T: William E. Klunk and Chester A. Mathis, "Whatever Happened to Pittsburgh Compound-A?" *Alzheimer Disease and Associated Disorders*, July–September 2008, pp. 198–203.

99 "Whether it was blind luck or the fruits of persistence, no one can say": Ibid.

100 Klunk and Mathis wrote a letter to that first, anonymous volunteer: Ibid.

100 PiB left the bloodstream, crossed into the brain, found plaques, and then worked its way into the center of the plaque: *The Forgetting*, 67:36.

100 The only thing that could hold them back now: Author's interviews with Chet Mathis and Bill Klunk.

101 They never got it: Ibid.

101 The discovery of PiB "represented a major breakthrough": Author's interview with David Kupfer.

102 In 2008, for their discovery of PiB: "Klunk and Mathis Win Prestigious Potamkin Prize For Alzheimer's Research," *Pitt Chronicle*, April 27, 2008.

TEN: DÉJÀ VU

103 Bill Klunk, who talked about his discovery of Pittsburgh Compound B: *The Forgetting*, 35:30.

103 They were ten siblings: Tanzi and Parson, *Decoding Darkness*, pp. 48, 98 (see chap. 3, n. 5).

103 Ultimately, she died of pneumonia: Kate Preskenis, "Family History," http://katepreskenis.com/family-history/.

103 One of her daughters, Julie Noonan Lawson, was interviewed: *The Forgetting*, 29:21.

104 "That's when we realized: We're not done": *The Forgetting*, Ibid.

104 In February 2004, Gail received: Author's interview with Karla DeMoe Hornstein, Aug. 17, 2011.

106 It was Alzheimer's disease: Author's interviews with Karla Hornstein, August 17, 2011, and April 5, 2013.

107 "I don't think anybody thought it would turn into this": Author's interview with Jennifer DeMoe, July 2012.

107 "I thought, 'Oh crap, *both* of them?'": Author's interview with Jessica McIntyre, August 15, 2012.

108 "I've been doing this for fifty years": Author's interview with Vic DeMoe, September 13, 2012.

109 Sunderland had arrived at the National Institute of Mental Health: David Willman, "$508,050 from Pfizer, but No 'Outside Positions to Note,'" *Los Angeles Times*, December 22, 2004.

109 Additional biographical details on Sunderland: David Willman, "Review Faults Scientist's Conduct," *Los Angeles Times*, September 10, 2006.

111 Medicare will not pay for such scans: "Decision Memo for Beta Amyloid Positron Emission Tomography in Dementia and Neurodegenerative Disease," Centers for Medicare & Medicaid Services, September 27, 2013.

ELEVEN: WHEN THE FOG ROLLS IN

113 She was just a tiny thing: Tanzi and Parson, *Decoding Darkness*, p. 84; Kosik and Clegg, *The Alzheimer's Solution*, p. 29 (see chap. 3, n. 3).

113 There was a time: Tanzi and Parson, *Decoding Darkness*, pp. 20, 48.

114 The ten children Julia left behind were: Author's interview with Julie Noonan Lawson, December 5, 2012.

114 When Kate Preskenis was in eighth grade: Author's interview with Kate Preskenis, November 13, 2012.

115 Fran's husband, a doctor, had attended a conference and returned with that suggestion: Author's email exchange with Julie Noonan Lawson, May 10–11, 2016.

115 Her request was emphatic: Kate Preskenis, *The Gene Guillotine: An Early-Onset Alzheimer's Memoir*, p. 15, http://katepreskenis.com/family-history/, accessed December 21, 2015.

117 Marilyn Albert: Author's interview with Marilyn Albert, February 21, 2014.

117 their age of onset was dramatically different: Preskenis, *The Gene Guillotine*.

117 She eventually died of lung cancer: Author's interview with Julie Noonan Lawson, December 5, 2012.

117 researchers later theorized the difference in their disease progression may

have been epigenetic: Author's email exchange with Ken Kosik, February 25, 2015; Andy Coghlan, "Hints of Epigenetic Role in Alzheimer's Disease," *New Scientist*, August 17, 2014; Mark J. Millan, "The Epigenetic Dimension of Alzheimer's Disease: Causal, Consequence, or Curiosity?" *Dialogues in Clinical Neuroscience*, September 16, 2014, pp. 373–393.

119 "I felt like I knew her": Author's interview with Karla Hornstein, January 23, 2012.

119 When Fran was still alive and well enough to speak: Congressional testimony of Frances Powers, "Breakthroughs in Brain Research: A National Strategy to Save Billions in Health Care Costs," hearing before the Senate Committee on Aging, US Senate, 104th Congress, June 27, 1995, serial no. 104–5, http://www.aging.senate.gov/imo/media/doc/publications/6271995.pdf.

TWELVE: *MALDICIÓN*

122 In pursuit of his enemy: Author's interview with Francisco Lopera, April 18, 2013.

122 He had been experiencing his symptoms for four years: Michael Jacobs, "Yarumal, Colombia: the Largest Population of Alzheimer's Sufferers," *Telegraph (UK)*, October 23, 2012.

123 *La bobera*, they called it: Pam Belluck, "A Perplexing Case Puts a Doctor on the Trail of 'Madness,'" *New York Times*, June 1, 2010.

123 almost two-thirds of Americans with Alzheimer's are women: "2015 Alzheimer's Disease Facts and Figures," Alzheimer's Association.

123 On one trip, while collecting blood samples in Angostura, Lucía was kidnapped by a drug cartel and held for eight days: Pam Belluck, "Alzheimer's Stalks a Colombian Family," *New York Times*, June 1, 2010.

124 During the years he studied them, he mapped out the family tree to extend to five thousand members, all descending from Javier San Pedro Gómez and María Luisa Chavarriaga Mejía: Belluck, "A Perplexing Case." Also author's interview with Francisco Lopera, April 18, 2013.

125 "I think the things that make people curious about the brain are all in literature": Author's interview with Ken Kosik, April 2, 2012.

127 "at a stage of life when very little prevents us from following our instincts and the allure of the unknown": Kosik and Clegg, *The Alzheimer's Solution*, p. 16 (see chap. 3, n. 3).

127 While Lopera conducted a basic neurological exam on a man in his late forties: Kosik and Clegg, *The Alzheimer's Solution*, pp. 16–17.

127 Kosik was moved by this world: Kosik and Clegg, *The Alzheimer's Solution*, p. 18.

128 Among Lopera's patients was a middle-aged mother of fourteen children: Belluck, "A Perplexing Case"; also Kenneth S. Kosik, "The Fortune Teller," *Sciences*, July/August 1999, p. 14.

129 the ravaged organs sat in round white plastic tubs: "The Remote Hope for Preventing Alzheimer's," Channel 4 News (UK), October 10, 2011, http://www.channel4.com/news/the-remote-hope-for-preventing-alzheimers.

130 But in Colombia, there were no genetic counselors: Kosik and Clegg, *The Alzheimer's Solution*, p. 25.

130 The only answer came from a twenty-three-year-old man named Gonzalez: Kosik, "The Fortune Teller," p. 16.

130 "We stand poised to be expelled from an Eden of genetic ignorance": Kosik, "The Fortune Teller," pp. 16–17.

131 homozygote: Author's interview with Ken Kosik and Francisco Lopera, April 18, 2013; Kenneth Kosik et al., "Homozygosity of the Autosomal Dominant Alzheimer's Disease Presnilin 1 E280A Mutation," *Neurology*, November 24, 2014.

THIRTEEN: BURDEN OF PROOF

135 "If I do have it": Author's interview with Robin Tjosvold, May 2012.

135 "We clicked": Author's interview with Lori McIntyre, May 4, 2012.

136 She had been visiting a friend in downstate Wisconsin: Recollection of Leah Klobucher and Alayna Alexson during a visit to the University of Pittsburgh Medical Center, January 2, 2013.

136 The Mayo Clinic reported: Ibid.

137 That matched the findings of a study that would be published in 2012 suggesting that patients taking Aricept would perform cognitively better: Robert Howard et al., "Donepezil and Memantine for Moderate-to-Severe Alzheimer's Disease," *New England Journal of Medicine*, March 8, 2012.

138 When Cole answered the phone: Author's interview with Cole Hornstein, February 13, 2014.

FOURTEEN: THE RISE AND FALL OF GOLDEN BOY

142 his office nickname was Golden Boy: Author's interview with Susan Molchan, February 6, 2014.

142 He won NIMH's Exemplary Psychiatrist Award: Biography of Trey Sunderland, MD, provided by the NIH Clinical Center for an event titled "Alzheimer's Disease: Advances and Hope" on September 16, 2003.

143 But there was another side to the Golden Boy: Author's interview with Susan Molchan, February 6, 2014.

143 To NIMH director Thomas Insel, the NIH was "Camelot": Congressional testimony of Thomas Insel, "Human Tissue Samples: NIH Research Policies and Practices," Subcommittee on Oversight and Investigations of the Committee on Energy and Commerce, US House of Representatives, 109th Congress, June 14, 2006, serial no. 109–119.

143 though theoretically there were boundaries to safeguard against any conflict of interest: Testimony of Thomas Insel, "Human Tissue Samples."

143 "The NIH is supposed to be above all that": Author's interview with Susan Molchan, February 6, 2014.

143 At the same time, a lot of government scientists were consulting for private companies: Testimony of Thomas Insel, "Human Tissue Samples."

144 Since Aricept's introduction: Katie Thomas, "Drug Dosage Was Approved Despite Warning," *New York Times*, March 22, 2012.

144 Cognex would be discontinued: Mona Mehta, Abdu Adem, and Marwan Sabbagh, "New Acetylcholinesterase Inhibitors for Alzheimer's Disease," *International Journal of Alzheimer's Disease*, November 7, 2011.

144 Looking to further the progress they'd just made with Aricept: Staff report written by the Majority and Minority Committee Staff of the House Committee on Energy and Commerce for the use of the Subcommittee on Oversight and Investigations in preparation for its hearing, "Human Tissue Samples: NIH Research Policies and Practices," June 13–14, 2006.

145 "the experience, and knowledge, and access to samples that would make this project possible": Testimony of David Friedman to the Subcommittee on Oversight and Investigations.

145 In total, Sunderland sent thirty-two hundred or so vials of plasma and spinal fluid to Pfizer: Ibid.

145 Pfizer wanted those vials to help develop drugs that would target the disease in its earliest stages: Testimony of David Friedman to the Subcommittee on Oversight and Investigations of the Committee on Energy and Commerce, June 14, 2006.

145 In April 2003, the storied *Journal of the American Medical Association* published some of their results: Trey Sunderland et al., "Decreased β-Amyloid$_{1\text{-}42}$ and Increased Tau Levels in Cerebrospinal Fluid of Patients With Alzheimer Disease," *Journal of the American Medical Association*, April 23, 2003.

145 But the company also used Sunderland to promote Aricept: Willman, "$508,050 from Pfizer" (see chap. 10, n. 11).

145 under which he was paid $25,000 per year as a consultant: Staff report written

by the Majority and Minority Committee Staff of the House Committee on Energy and Commerce for the use of the Subcommittee on Oversight and Investigations.

145 All told, he earned roughly $500,000 from his five-year private arrangement: Ibid.

145 Of that amount, Sunderland failed to disclose about $300,000 to his bosses: David Willman, "Illicit payments from Pfizer Cost NIH Researcher $300,000," *Los Angeles Times*, December 23, 2006; John T. Papavasiliou, Deputy Director, Maryland State Board of Physicians, Final Decision and Order, March 30, 2009, Case Number 2007-0392.

145 From 1999 to June 2004, he took the show on the road: Willman, "$508,050 from Pfizer."

146 Molchan was a graduate: Author's interview with Susan Molchan, February 6, 2014.

146 Molchan embarked on a study that tested whether lithium—an element sometimes used to treat mania—might help prevent tau proteins from becoming toxic: Author's interview with Susan Molchan, March 3, 2014.

146 However, Molchan took more fluid than was typical, with the intent of storing some for future research: Staff report written by the Majority and Minority Committee Staff of the House Committee on Energy and Commerce for the use of the Subcommittee on Oversight and Investigations.

146 Molchan began to think that her studies got less support: Author's interview with Susan Molchan, February 6, 2014.

147 Sunderland promised to support her for a tenure-track position: Author's interview with Susan Molchan, February 6, 2014.

147 Sunderland informally diagnosed her as paranoid and depressed, while another woman was ordered to see a psychiatrist: Author's interview with Susan Molchan, February 6, 2014; complaint in the Equal Employment Opportunity Commission (EEOC) case of Susan Molchan.

147 she filed a sex discrimination claim that was dismissed: EEOC decision in the case of Susan Molchan v. Donna E. Shalala, Appeal No. 01982167, October 29, 1998.

147 she was alerted to a lithium study requiring spinal fluid similar to the archive she'd left behind at NIMH: Staff report written by the Majority and Minority Committee Staff of the House Committee on Energy and Commerce for use of the Subcommittee on Oversight and Investigations.

147 "Now, I understand we didn't need a whole lot": Testimony of Susan Molchan to the Subcommittee on Oversight and Investigations, June 13, 2006.

147 But when she asked what happened to the rest, Sunderland appeared to hedge: Ibid.

147 True, at the time, the NIH's policies for tracking its human tissue samples were erratic: Testimony of Thomas Insel, "Human Tissue Samples."

148 However, for Molchan this atmosphere of carelessness didn't explain away Sunderland's vagueness about the vials: Testimony of Susan Molchan to the Subcommittee on Oversight and Investigations.

148 she contacted the inspector general and the Department of Health and Human Services: Prepared statement of Ed Whitfield, chairman, Subcommittee on Oversight and Investigations, June 13, 2006.

150 "Not disclosing over $500,000 [*sic*] in income": Staff report written by the Majority and Minority Committee Staff of the House Committee on Energy and Commerce.

150 "I thought of Dr. Sunderland as one of the people who had made tremendous contributions": Testimony of Thomas Insel, "Human Tissue Samples."

150 "What little information that we have gotten, some of it appears to be misleading or intentionally inaccurate": Response by US Representative Joe Barton to Thomas Insel, Subcommittee on Oversight and Investigations hearing, June 14, 2006.

150 Insel pointed out that the ethics case had been referred to the Department of Justice: Testimony of Thomas Insel, "Human Tissue Samples."

150 "We made a referral": Ibid.

151 Barton was sympathetic, but only to a point: Response of US Representative Joe Barton to Thomas Insel, Subcommittee on Oversight and Investigations hearing, June 14, 2006.

151 Trey Sunderland pleaded guilty: "NIH Scientist Sunderland Pleads Guilty to Conflict-of-Interest Charge," *FDA News*, December 12, 2006; David Willman, "Illicit Payments from Pfizer Cost NIH Researcher $300,000," *Los Angeles Times*, December 23, 2006.

151 "This process has humbled me": David Willman, "Illicit payments from Pfizer."

151 Despite her protests against his use of her samples, Molchan remained sympathetic to her former boss on a personal level: Author's interview with Susan Molchan, February 6, 2014.

152 His medical license was revoked in Maryland in 2009 and in New York in 2011: John T. Papavasiliou, Deputy Director, Maryland State Board of Physicians, Final Decision and Order, March 30, 2009, Case Number 2007-0392; Kendrick A. Sears, Chair, State Board for Professional Medical Conduct, New York State Department of Health, Consent order BPMC: 11-10, January 11, 2011.

152 "He has not convinced the Board": John T. Papavasiliou, Deputy Director, Maryland State Board of Physicians, Board Decision and Order on Petition for Reinstatement, Case Number 2007-0392, December 20, 2010.

152 One, who joined the research because his own father had died with Alzheimer's: Willman, "Illicit payments from Pfizer."

152 Julie Noonan Lawson was deeply saddened: Author's interview with Julie Noonan Lawson, February 12, 2014.

153 At the time of the investigation, a scientist in Sunderland's position: Testimony of Thomas Insel, "Human Tissue Samples."

153 But what bothered Lawson the most: Author's interview with Julie Noonan Lawson, February 12, 2014.

153 The information and samples he collected were warehoused, and over time, some of them were lost: Author's interview with Marilyn Albert, February 21, 2014.

153 Albert was awarded $1.5 million: Ibid.

153 In return, she received: Ibid.

154 Karla, who was devastated: Author's email exchange with Karla Hornstein, December 6, 2012.

154 One of the most poignant losses in the Sunderland affair: Author's interview with Julie Noonan Lawson, February 12, 2014.

154 Marilyn Albert tried to help: Author's interview with Marilyn Albert, February 21, 2014.

154 Documentation was important to Marilyn Albert: Ibid.

FIFTEEN: FAVORITE SON

156 Debbie Thompson, Brian's high-school girlfriend: Author's interview with Debbie Ness, April 10, 2014.

157 Brian was living in a trailer on the outskirts of Tioga: Author's interview with Karla Hornstein, March 2, 2014.

157 Kassie found it ironic: Author's interview with Kassie Rose, August 23, 2011.

158 In more advanced cases of dementia: "Traveling with Dementia," Alzheimer's Association, http://www.alz.org/care/alzheimers-dementia-and-traveling.asp.

158 Karla tried to interest the Oklahoma cousins: Author's interviews with Karla Hornstein, January 23, 2012, and Sharon DeMoe, December 3, 2012.

159 "There's only one family like the DeMoes": Author's interview with William Klunk, September 12, 2011.

159 At the local drive-in, family friend Kim Johnston: Author's interview with Kim Johnston, April 7, 2014.

160 he forgot to let go of the fishing line and nearly lost a finger: Author's interview with Karla Hornstein, August 17, 2011.

160 The answer came soon enough, in February 2008: Author's interview with Karla Hornstein, March 2, 2014.

162 Brian's son, Yancey, later got rid of the cat: Author's interview with Kassie Rose, August 23, 2011.

162 When he discovered that he wasn't free to leave, he became enraged at Karla: Author's interview with Karla Hornstein, March 2, 2014.

163 "He [was] still part of my life because of my children": Author's interview with Christy Holm, July 2012.

163 On one visit, a friend left the building to light a cigarette: Author's interview with Karla Hornstein, March 2, 2014.

163 "I made a choice to live my life as normally as possible and leave it up to God": Author's interview with Kassie Rose, April 16, 2009.

163 "You know, you're just carrying that huge weight on your shoulders": Author's interview with Yancey DeMoe, September 2, 2012.

164 But families of Alzheimer's patients often look for clues to their future: Kosik and Clegg, *The Alzheimer's Solution*, p. 20 (see chap. 3, n. 3).

164 "Hey," Brian said. "I know you": Author's interview with Yancey DeMoe.

165 "I kind of knew my secret was safe with him": Author's interview with Kassie Rose, August 23, 2011.

SIXTEEN: THE BAPTISTS AND THE TAUISTS

169 The human body produces the amyloid precursor protein (APP) in several organs: "APP: Amyloid Beta Precursor Protein," *Genetics Home Reference*, US National Library of Medicine, reviewed May 2012, published May 10, 2016. https://ghr.nlm.nih.gov/gene/APP.

169 In Alzheimer's patients, something goes awry in the process: Shenk, *The Forgetting*, pp. 143–145 (see chap. 1, n. 7).

170 In the amyloid-versus-tau debate, Paul Aisen: Author's interview with Paul Aisen, November 26, 2013.

170 For years, he was fascinated with the earliest stages of Alzheimer's: Ibid.

171 Allen Roses of Duke University, who was famous for his 1992 discovery of apolipoprotein (ApoE) gene variants: Shenk, *The Forgetting*, p. 153; Natalie Angier, "Scientists Propose Novel Explanation for Alzheimer's," *New York Times*, November 9, 1993; Turna Ray, "Do We All Have Alzheimer's Completely Wrong? This Man Says Yes," Public Radio International, May 3, 2015.

171 "I can go into any cemetery and find a tombstone over a dead person": Author's interview with Eric Reiman, October 2012; Robert Finn, "Neuroscience Meeting to Feature Feisty Debate on Alzheimer's Etiology," *Scientist*, October 16, 1995.

171 ApoE is a gene that codes a protein: Topic sheet, "Genetic Testing," Alzheimer's Association.

171 The theory says E2 and E3 bind to tau: Angier, "Scientists Propose Novel Explanation."

171 In addition to not stabilizing tau, they say it also appears to promote the buildup of beta-amyloid: Troy T. Rohn, "Proteolytic Cleavage of Apolipoprotein E4 as the Keystone for the Heightened Risk Associated with Alzheimer's Disease," *International Journal of Molecular Sciences*, July 14, 2014; S. Ye, "Apolipoprotein (Apo) E4 Enhances Amyloid Beta Peptide Production in Cultured Neuronal Cells: Apoe Structure as a Potential Therapeutic Target," *Proceedings of the National Academy of Sciences of the United States of America*, December 20, 2005.

171 Reiman frames the amyloid-versus-tau debate with a colorful analogy: Author's interview with Eric Reiman.

172 In 1988, he joined the local Alzheimer's Association chapter in Phoenix, Arizona: Ibid.

172 he read a story in the *Wall Street Journal* about the discovery of the ApoE connection: Ibid.

172 Glucose, the main sugar in the blood, is the primary fuel that the brain uses for energy: Chun-Ling Dai, "Role of O-GlcNAcylation in Tau Pathology and Cognitive Function," Research Foundation for Mental Hygiene, Inc., at New York State Institute for Basic Research Staten Island, New York, 2015; Samuel Henderson, "Targeting Diminished Cerebral Glucose Metabolism for Alzheimer's Disease," *Drug Discovery World*, summer 2013.

172 although some researchers believe impaired glucose uptake helped to form the tau tangles: Ying Liu et al, "Decreased glucose transporters correlate to abnormal hyperphosphorylation of tau in Alzheimer disease," *ScienceDirect*, January 2008.

173 Reiman thought he could look at a couple of hundred E4 carriers over two years: Author's interview with Eric Reiman.

173 "Over half of family caregivers become clinically depressed": Ibid.

173 which often does not fully cover the cost of patient care: Cynthia Ramnarace, "The High Cost of Caring for Alzheimer's Patients," *AARP Bulletin*, October 18, 2010.

174 they still wouldn't win FDA approval solely on that basis: Author's interview with Eric Reiman.

174 Recent medical history was littered with examples: Ibid.

174 The FDA has a system in place: Fact sheet, "Accelerated Approval," US Food and Drug Administration, updated September 15, 2014.

174 A British colleague sitting next to Reiman: Author's interview with Eric Reiman.

175 Already, there were hints that the government would be open to that kind of flexibility: Ibid.

175 At the University of Rochester, Tariot built: Author's interview with Pierre Tariot, October 2012.

SEVENTEEN: EXPELLED FROM EDEN

178 Though the decision to know her results was terrifying: Author's interview with Jennifer DeMoe, July 2012.

178 even so, Alzheimer's was one topic they rarely discussed: Author's interview with Sheryl Grammer, January 24, 2012.

178 Their mother, Sharon, worried privately: Author's interview with Sharon Bratton DeMoe, December 3, 2012.

178 her grandfather Rob and aunt bluntly told her she was making a mistake: Author's interview with Leah Klobucher, January 2, 2013; author's interview with Robin Harvey, August 30, 2012.

179 He didn't see the value of Leah burdening herself with the disease: Author's interview with Jason Klobucher, January 2, 2013.

179 Her father, a Type I diabetic since childhood, agreed: Author's interview with Reed Alexson, January 31, 2013.

180 Jessica, Steve and Lori's oldest daughter, immediately made plans: Author's interview with Jessica McIntyre, August 15, 2012.

181 Dean, the brother he'd always emulated most, sat Jamie down for a heart-to-heart talk: Author's interviews with Dean DeMoe, Chelsey Determan, July 2012.

182 Though Jamie had been stunned, Karla thought he took the news surprisingly well: Author's interview with Karla Hornstein.

183 "The generation of six": Author's interview with Bill Klunk, September 12, 2011.

183 Chelsey knew Jamie DeMoe by reputation: Author's interview with Chelsey Determan, July 2012.

185 A surgeon reconstructed his face: Author's interview with Karla Hornstein.

185 The next time he went to Pittsburgh for his research visit, his results showed his cognitive decline had worsened: HBO documentary, *The Alzheimer's Project: The Supplementary Series.*

186 The days were long for Doug and Gail: Author's interview with Gail DeMoe, August 24, 2011.

186 Like Gail, Gary searched for ways to occupy his friend: Author's interview with Gary Anderson, October 20, 2015.

186 Since Karla held power of attorney for her brother: Author's interview with Karla Hornstein.

187 She called Steve: Author's interviews with Karla Hornstein, Steve McIntyre.

187 When they drove back home from Tioga, Deb would often suggest: Author's interview with Deb DeMoe, March 16, 2012.

189 Dean asked his closest friend, Monte Olson, for the same favor: Author's interview with Monte Olson.

189 but it is a fairly common one among Alzheimer's patients: Russell Powell, "Is Preventive Suicide a Rational Response to a Presymptomatic Diagnosis of Dementia?" *Journal of Medical Ethics* 40, issue 8, August 2014, pp. 511–512; Kosik and Clegg, *The Alzheimer's Solution*, pp. 195–197 (see chap. 3, n. 3).

189 tests predicting the likelihood of Alzheimer's are so rarely permitted outside of research studies: "Alzheimer's Disease Genetics Fact Sheet—Genetic Testing," Alzheimer's Disease Education and Referral Center, National Institute on Aging.

189 "The looming prospect of identity annihilation": Powell, "Is Preventive Suicide a Rational Response."

189 dementia raises a specific legal dilemma: Ibid.

191 The death of her father prompted Kassie to begin rethinking her decision to remain ignorant of her status: Author's interview with Kassie Rose, August 23, 2011.

EIGHTEEN: SAFE HAVENS

193 The solution was hydraulic fracturing: "The Bakken Boom: An Introduction to North Dakota's Shale Oil," Energy Policy Research Foundation, Inc., August 3, 2011.

194 In 2001, less than 2 percent of North Dakota's oil came from the Bakken; in 2011, more than 80 percent of it did: James Vlahos, "Oil Boom: North Dakota Is the Next Hub of U.S. Energy," *Popular Mechanics*, June 13, 2012.

194 he became the best tool pusher some of the company's men had ever seen: Author's interview with Rikki Rice, July 2012.

195 From 2005 to 2011, violent crime in the Williston Basin: Sari Horwitz, "Dark Side of the Boom," *Washington Post*, September 28, 2014.

196 The Alzheimer's Association reports that more than 60 percent of dementia patients wander: "Three Out of Five People with Alzheimer's Disease Will Wander," Alzheimer's Association of Northern California and Northern Nevada, http://www.alz.org/norcal/in_my_community_18411.asp.

196 Adding to the problem is the fact that Alzheimer's patients sometimes think they are in a different time in their lives: Kirk Johnson, "More With Dementia Wander from Home," *New York Times*, May 4, 2010.

196 At first, the home's staff demurred: Author's interview with Karla Hornstein.

197 it pained Gary to see his wild, reckless childhood buddy so changed: Author's interview with Gary Anderson.

NINETEEN: A BIG IF

199 Tariot began taking off whole days: Author's interview with Pierre Tariot, October 24, 2012.

200 Reiman's and Tariot's cognitive tests were designed to work on patients as young as thirty: Ibid.

200 Among them was Don Berry, the lead statistician for the MD Anderson Cancer Center at the University of Texas: Author's interviews with Eric Reiman and Pierre Tariot, October 24, 2012.

200 Though the meeting was not formal or legally binding, both agencies indicated that Banner was moving in the right direction: Author's interview with Pierre Tariot, October 24, 2012.

201 the Alzheimer's Association was publicly predicting that one in eight of them would develop the disease: "Generation Alzheimer's: The Defining Disease of the Baby Boomers," Alzheimer's Association, 2011.

201 In Colombia, Lopera was in the middle of discussions with Swiss pharmaceutical giant Novartis: Author's interviews with Francisco Lopera and Ken Kosik, April 18, 2013.

202 The Colombian team wanted to work in what is known as a pre-competitive atmosphere: Ibid.

202 When Jonas Salk announced that he'd created a polio vaccine: Jon Cohen, *Shots in the Dark: The Wayward Search for an AIDS Vaccine* (New York: W. W. Norton, 2001), 88.

202 But years before him, a virologist named Hilary Koprowski mixed his own vaccine: Margalit Fox, "Hilary Koprowski, Who Developed First Live-Virus Polio Vaccine, Dies at 96," *New York Times*, April 20, 2013.

203 *curanderos*—traditional Latin American shamans—practice alongside doctors: Author's interview with Ken Kosik, April 18, 2013.

203 "We think your sons are heroes," he said: CNN special, *Filling In the Blanks*.

203 "You cannot meet these families and not be transformed," said Reiman: Author's interview with Eric Reiman, October 24, 2012.

TWENTY: EVERYONE SEES THE POWER

206 he launched an international study: Author's interview with John Morris, March 20, 2012.

208 "It's hard on that old lady": Author's interview with Yancey DeMoe, September 2, 2012.

208 One of their rooms was filled with documents: Author's interview with Matt Hornstein, July 2012.

209 When a psychiatrist interviewed Jamie about his life: Jamie DeMoe, visit to the University of Pittsburgh, September 12, 2012.

209 Depression is extremely common in Alzheimer's patients: "Depression and Alzheimer's," Alzheimer's and Dementia Caregiver Center, Alzheimer's Association.

210 The National Institute of Mental Health established formal guidelines: Ibid.

210 "if you ask Jamie, he always says three": Author's interview with Chelsey Determan, September 2012.

212 As symptoms advance, Alzheimer's patients struggle more with travel: "Traveling with Dementia," Alzheimer's and Dementia Caregiver Center, Alzheimer's Association.

212 "You want to know why I do it?": Author's interview with Jamie DeMoe, September 12, 2012.

212 "All he ever says about Savannah is that he hopes to see her graduate": Author's interview with Rikki Rice, July 2012.

TWENTY-ONE: LANDSLIDE

214 "What's wrong with twenty-five great years?": Author's interview with Reed Alexson, January 31, 2013.

215 She was saddened by Dawn's death: Author's interview with Robin Harvey, October 7, 2013.

215 On a late July weekend in 2004: Accounts of the accident that killed Mike Harvey taken from author's interviews with Robin Harvey, August 30, 2012; author's interview with Taylore Ogren, August 30, 2012; Shelley Nelson,

"Event to Benefit Children Whose Father Died Trying to Save Them," *Duluth News Tribune*, October 1, 2004; Shelley Nelson, "Two Drown in Attempted Rescue," *Duluth News Tribune*, July 27, 2004; author's interview with rescuer Phil Anderson, July 13, 2014.

215 She was furious with Leah: Author's interview with Robin Harvey, August 30, 2012.

216 The psychiatrist asked Robin: Robin Harvey's visit to the University of Pittsburgh, August 29, 2012.

216 Her personality was an odd mashup: Author's interview with Colleen Miller, December 12, 2012.

217 "She's got the kindest heart": Author's interview with Becky Vork, December 13, 2012.

219 "If I think about it too much": Ibid.

TWENTY-TWO: SOMETHING TO SHOOT FOR

220 Some sources peg the median cost: Matthew Herper, "The Cost of Creating a New Drug Now $5 Billion, Pushing Big Pharma to Change," *Forbes*, August 11, 2013.

220 In the early 1980s, drugmaker Parke-Davis: Ron Winslow, "The Birth of a Blockbuster: Lipitor's Route Out of the Lab," *Wall Street Journal*, January 24, 2000; Andrew Jack, "The Fall of the World's Best-Selling Drug," *Financial Times*, November 28, 2009.

222 But as additional studies began to back up the tau theory, suggesting it was the primary driver, drug development was following suit: Cynthia Fox, "Tau Is the Main Culprit in Alzheimer's, Say Two Studies," *Drug Discovery & Development*, April 7, 2015; Jim Schnabel, "Target: Tau," the Dana Foundation, September 23, 2014.

222 as was the push to image tau proteins the way PiB imaged amyloid: Author's interview with Reisa Sperling, November 5, 2013.

222 Some scientists, including Eric Reiman: Author's interview with Eric Reiman, October 2012.

222 From 1998 to 2011, a whopping 101 Alzheimer's treatments had failed to reach patients: "From Setbacks to Stepping Stones: Alzheimer's Disease," The Pharmaceutical Research and Manufacturers of America press release, September 13, 2012.

223 one that delayed onset by even five years would be considered a major victory: Author's interview with Eric Reiman, October 2012.

223 A 2015 report compiled by the Dementia Forum of the World Innovation

Summit for Health: Ellis Rubinstein et al., "A Call to Action: The Global Response to Dementia Through Policy Innovation," report of the WISH Dementia Forum 2015.

223 The committee charged with selecting drugs: The selection of DIAN's trial drugs and the study protocol came from several sources, including: Gabrielle Strobel, "2012 DIAN Series," Alzheimer's Research Forum; "Dominantly Inherited Alzheimer Network Trial: An Opportunity to Prevent Dementia. A Study of Potential Disease Modifying Treatments in Individuals at Risk for or with a Type of Early Onset Alzheimer's Disease Caused by a Genetic Mutation," description of the study's outcome measures, ClinicalTrials.gov, a service of the US National Institutes of Health; author's interview with Randy Bateman, June 17, 2013.

223 But data analysis from those trials suggested that it slowed: Results on solanezumab drawn from Gwyneth Dickey Zakaib, "Phase 3 Solanezumab Trials 'Fail'—Is There a Silver Lining?" Alzheimer's Research Forum, August 24, 2012; Charlotte Jago: "Alzheimer's Disease: One Year Later," *Life Sciences Connect*, February 3, 2014.

224 In cases like Dawn's daughters, Alayna and Leah, determining their mother's age of onset was difficult: Visit of Leah Klobucher and Alayna Alexson to the University of Pittsburgh, January 2, 2013.

226 In Oklahoma, forty-two-year-old Sherry DeMoe Pickard: Author's interview with Sherry Pickard, April 8, 2014.

226 "There's times when I think, 'I can't do this again'": Author's interview with Sharon DeMoe, December 3, 2012.

TWENTY-THREE: THE SILVER TSUNAMI

232 For example, it successfully debunked the myth that ginkgo biloba: Elaine Woo, "Dr. Leon Thal, 62, UC San Diego Alzheimer's Expert Killed in Crash," *Los Angeles Times*, Februrary 8, 2007; for further information on ginkgo biloba's lack of effect on cognitive decline, see also: Beth E. Snitz et al., "Ginkgo biloba for Preventing Cognitive Decline in Older Adults: A Randomized Trial," *Journal of the American Medical Association*, December 23/30, 2009.

232 Solanezumab carried enough safety data to convince Reisa Sperling: Author's interview with Reisa Sperling, November 5, 2013.

232 After all, some people carried amyloid in their brains, yet functioned normally: Reisa Sperling, "Can We Treat Alzheimer's 20 Years Early?" TEDMED, video presentation published July 17, 2012.

232 Several reports about those concerns: Anne M. Summer, "The Silver Tsunami: One Educational Strategy for Preparing to Meet America's Next Great Wave of Underserved," *Journal of Health Care for the Poor and Underserved*, August 2007; "The Silver Tsunami," *Economist*, February 4, 2010; Amy Burkholder, "Alzheimer's and the 'Silver Tsunami': Is America Ready?" CBS News, December 14, 2010.

232 In 2013, a study published in the *New England Journal of Medicine* predicted: Michael D. Hurd et al., "Monetary Costs of Dementia in the United States," *New England Journal of Medicine*, April 4, 2013.

232 They shared information with Aisen and Sperling: Author's interview with Paul Aisen, January 15, 2016.

233 which affects about 2 percent of the general population: Derek Lowe, "ApoE4: Test or Not?" In the Pipeline, *Science Translational Medicine*, February 28, 2008; author's interview with Eric Reiman, October 2012; author's interview with Reisa Sperling; "2016 Alzheimer's Disease Facts and Figures," Alzheimer's Association, p. 11.

233 Researchers initially believed carrying two copies of ApoE4 represented a 90 percent chance: Author's interviews with Pierre Tariot, October 2012 and June 2, 2016; E. H. Corder et al., "Gene Dose of Apolipoprotein E Type 4 Allele and the Risk of Alzheimer's Disease in Late Onset Families," *Science*, August 13, 1993.

233 subsequent risk estimates have lowered that number to 58 to 68 percent: Due to the inherent limitations of statistical techniques, scientists typically express genetic risk in a range, according to Pierre Tariot. The most recent risk percentages associated with two copies of ApoE4 were drawn from the author's interview with Pierre Tariot, June 2, 2016; estimates by 23andMe, a direct-to-consumer genetic testing company, and the Risk Evaluation and Education for Alzheimer's Disease (REVEAL) Study, a multicenter trial examining the impact of genetic testing and disclosure. (For more details about REVEAL, see: Serena Chao et al., "Health Behavior Changes After Genetic Risk Assessment for Alzheimer Disease: The REVEAL Study," *Alzheimer Disease & Associated Disorders*, 2008.

233 about the same risk that women with the BRCA1 gene mutation: Genin et al., "APOE and Alzheimer Disease: A Major Gene with Semi-Dominant Inheritance," *Molecular Psychiatry*, May 10, 2011.

233 including actress Angelina Jolie: Angelina Jolie, "My Medical Choice," *New York Times*, May 14, 2013.

233 roughly 30 to 45 percent: Table of risk estimates from 23andMe and the

REVEAL study, ranges interpreted by Pierre Tariot. The report "2016 Alzheimer's Disease Facts and Figures," from the Alzheimer's Association, p. 11, estimates that 23 percent of the general population carries one copy of the ApoE4 gene variant.

233 The A4 study hoped to enroll a thousand people: Author's interview with Reisa Sperling.

234 To approve a medication for use in Alzheimer's patients: Details about the FDA's revised approach to approving drugs for Alzheimer's Disease can be found in the following: "Guidance for Industry—Alzheimer's Disease: Developing Drugs for the Treatment of Early Stage Disease," US Department of Health and Human Services, Food and Drug Administration, Center for Drug Evaluation and Research (CDER), February 2013; Nicholas Kozauer and Russell Katz, "Regulatory Innovation and Drug Development for Early-Stage Alzheimer's Disease," *New England Journal of Medicine*, March 28, 2013; Gina Kolata, "F.D.A. Plans Looser Rules on Approving Alzheimer's Drugs," *New York Times*, March 13, 2013.

234 But a surprising answer came out of the University of California, San Francisco: Author's interview with Paul Aisen, November 26, 2013.

235 The Mini Mental State Examination, known to clinicians as the MMSE: Ibid.

235 The first person ever dosed with PiB scored 25: William E. Klunk and Chester A. Mathis, "Whatever Happened to Pittsburgh Compound-A?" *Alzheimer Disease & Associated Disorders*, July–September 2008.

235 But the MMSE is so imperfect that even people with dementia can sometimes score a perfect 30: Author's interview with Paul Aisen, November 26, 2013.

235 Julie Noonan Lawson thought the test: Author's interview with Julie Noonan Lawson, December 5, 2012.

236 The FDA also agreed to make its rules more flexible, thanks in large part to the efforts of Russell "Rusty" Katz: Author's interview with Eric Reiman and Pierre Tariot, October 2012; author's interview with Paul Aisen, November 26, 2013.

236 Reisa Sperling wasn't expecting a home run from A4: Author's interview with Reisa Sperling.

237 "Amyloid pulls the trigger, but tau is the bullet": Kay Lazar, "Researchers Seek to Short-Circuit Alzheimer's," *Boston Globe*, March 27, 2015.

237 To help answer those questions, they added yet another resource to their study: Author's interview with Reisa Sperling.

237 For example, in November 2013, he told colleagues: Madolyn Bowman Rogers, "Do Tau Tracers Track Cognitive Decline in Disease?" Alzheimer's Research Forum, November 27, 2013; press release from the Clinical Trials

on Alzheimer's Disease, "Tau in the Brain: A Better Predictor of Alzheimer's Disease Progression?" November 14, 2013.

TWENTY-FOUR: THE LUCKY ONES

240 In the summer of 2008, twenty-three-year-old Chelsey McIntyre, Steve and Lori's youngest, was facing the toughest dilemma of her life: Author's interview with Chelsey McIntyre, June 7, 2012.

241 both her parents were adamantly opposed to any of their daughters finding out their status: Author's interviews with Steve and Lori McIntyre.

241 "For right now, I'm perfectly content not knowing": Author's interview with Lindsey Sillerud: September 14, 2012.

242 Deb DeMoe, who turned so often to her faith for answers, consulted her pastor: Author's interview with Deb DeMoe, March 16, 2012.

242 "It was the second or third [diagnosis] when I started to pick up that wow, this was serious": Author's interview with Amber Hornstein, July 2012.

243 "She's always thinking about it, always talking about it": Author's interview with Matt Hornstein, July 2012.

243 "She's never going to lighten her load": Author's interview with Cole Hornstein, February 13, 2014.

243 One night at Grandma Gail's house in Tioga: Author's interviews with Amber Hornstein, July 2012, and Chelsey McIntyre, June 2012.

244 It was then that Karla realized Sherry had the disease: Author's interview with Karla Hornstein, January 23, 2012.

TWENTY-FIVE: FOLLOW THE SCIENCE

245 about fifty of the Colombian *paisa* were traveling in groups to the Banner Institute: Author's interview with Pierre Tariot, October 2012.

245 Eleven of the *paisa* already had symptoms: Details of the *paisa* visit to the Banner Institute were drawn from Adam S. Fleisher et al., "Florbetapir PET Analysis of Amyloid-ŒÍ Deposition in the Presenilin 1 E280A Autosomal Dominant Alzheimer's Disease Kindred: A Cross-Sectional Study," *Lancet*, November 6, 2012.

246 And so it came to be that Pierre Tariot and his wife, Laura, welcomed groups of the *paisa*: Author's interview with Pierre Tariot, October 2012.

247 In the *paisa*, amyloid begins increasing at the age of twenty-eight: Author's interview with Eric Reiman, October 2012.

247 "This is a special population": Ibid.

248 Tariot and Reiman invited fellow Alzheimer's researchers from around the

country: Author's interview with Eric Reiman and Pierre Tariot, October 2012.

249 Crenezumab began its life as a molecule: Author's interview with Andrea Pfeifer, October 21, 2013.

250 Carole Ho, the neurologist who oversees Genentech's early clinical development group: Author's interview with Carole Ho, November 13, 2013.

250 To Tariot, the company's decision was nothing short of brilliant: Author's interview with Eric Reiman and Pierre Tariot, October 2012.

251 they sought the advice of Jur Strobos: Ibid.

251 "We think we have a chance": Author's interview with Eric Reiman, October 2012.

252 On December 23, 2013, the Banner Alzheimer's Institute announced: Press release, Banner Alzheimer's Institute: "Alzheimer's Prevention Initiative Marks Milestone," December 23, 2013.

252 "We don't have time to do all the research we can do": Author's interview with Francisco Lopera and Ken Kosik, April 18, 2013.

TWENTY-SIX: LIKE FATHER, LIKE SON

253 After consulting with Dean and another friend, he turned it down: Author's interview with Monte Olsen.

254 He visited, but only every so often; and when he did, it ruined his day: Author's interview with Dean DeMoe, October 13, 2015.

255 In Santa Barbara, California, Ken Kosik's university research team thought it might have one answer: Julie Cohen, "Studying the Outliers," *UC Santa Barbara Current*, September 1, 2015; Mimi Liu, "UCSB Researchers Find Possible Factor to Delay Onset of Alzheimer's," *Daily Nexus*, October 1, 2015; Geoffrey Riley, "Taking Steps Now to Avoid Alzheimer's," Jefferson Public Radio, February 1, 2016; author's interview with Ken Kosik, March 2, 2015.

255 Tyler thanked him for the opportunity: Research visit by Tyler DeMoe to the University of Pittsburgh, June 11, 2012.

257 After talking to a genetic counselor: Author's interview with Tyler DeMoe, June 12, 2012.

TWENTY-SEVEN: ALL THE CARDS ARE ON THE TABLE

258 Seated at a conference table: Visit of Lori McIntyre to the University of Pittsburgh, May 4, 2012.

258 "I threw my hardship card on the table": Author's interview with Steve McIntyre, May 4, 2012.

259 As a nurse, Robin Tjosvold had worked in settings: Author's interview with Robin Tjosvold, May 2012.

260 "He's taking the whole thing like a champ": Author's interview with Jessica McIntyre, August 15, 2012.

261 To Steve, it wasn't just the gradual loss of his wife that stung: Author's interview with Steve McIntyre, May 2012.

264 But Lori said the silence of her friends was understandable: Author's interview with Lori McIntyre, May 2012.

265 she and Steve sat on a conference call with Bill Klunk: Author's notes from conference call, November 9, 2012.

TWENTY-EIGHT: COMING HOME

269 Steve wanted to hire a caregiver: Author's interview with Steve McIntyre, November 26, 2013.

270 Robin drove to her parents' house: Author's interview with Robin McIntyre, November 14, 2013.

270 That fall, when Steve had to be hospitalized with pancreatitis: Author's interview with Steve McIntyre, November 26, 2013.

271 In their phone call: Author's interview with Karla Hornstein, November 20, 2013.

272 The day after Thanksgiving 2013: Accounts of Lori's move into the nursing home compiled from author's interviews with Steve McIntyre and Karla Hornstein.

272 Bill Klunk volunteered to help: Author's interview with Steve McIntyre, February 21, 2014.

273 There is a curious symmetry to Alzheimer's: "Stages of Alzheimer's," Alzheimer's Association, http://www.alz.org/alzheimers_disease_stages_of_alzheimers.asp?type=brainTourFooter.

TWENTY-NINE: PALPABLE MOMENTUM

274 in May 2012, the Obama administration named Alzheimer's disease: National Alzheimer's Plan 2012, p. 3.

274 "The question is, can we really beat this disease?": Author's interview with Randy Bateman, January 12, 2016.

275 By comparison, the NIH spends $3 billion a year on AIDS research: Monica Brady-Myerov, "Alzheimer's Funding Lags Behind Other Diseases," WBUR, October 20, 2011; "Generation Alzheimer's: The Defining Disease of the Baby Boomers," Alzheimer's Association, 2011.

275 The Alzheimer's Association reported that in 2013: "New Alzheimer's Association Report Reveals 1 in 3 Seniors Dies with Alzheimer's or Another Dementia," press release, Alzheimer's Association, March 19, 2013.

275 Bateman's DIAN trial: Press release, "Alzheimer's Association Awards Largest Ever Research Grant to the Dominantly Inherited Alzheimer's Network (DIAN) for Innovative Therapy Trials," March 20, 2012.

275 "If you had asked me five years ago": Gabrielle Strobel, "DIAN Forms Pharma Consortium, Submits Treatment Trial Grant," Alzheimer Research Forum, December 22, 2011.

276 But the DIAN team had a significant scare at the end of 2012: Author's interview with Randy Bateman, June 17, 2013.

277 Dr. Francis Collins, director of the NIH: "NIH's Collins Delivers Positive News About Alzheimer's Research Funding," Alzheimer's Association Advocacy Forum, April 23, 2013.

THIRTY: TO THE MOON AND BACK

278 On Friday evening, she spoke with her sister: Author's interview with Deb DeMoe, June 8, 2013.

279 On Saturday, Dean headed to the other side of the state: Ibid.

279 Gail offered to babysit six-year-old Savannah: Author's interview with Chelsey Determan, October 7, 2013.

281 Gail took a few steps onto the concrete sidewalk: Author's interviews with Chelsey Determan and Savannah DeMoe, October 7, 2013; author's interview with Deb DeMoe, June 8, 2013.

283 It was the best family reunion Robin McIntyre had ever attended: Author's interview with Robin McIntyre.

284 A week later, when Karla was driving home: Author's interview with Karla Hornstein.

285 The following year, on May 27, 2015: Author's interviews with Deb DeMoe, January 23, 2016, and February 8, 2016. Confirmed by Tamara Donahue, research nurse coordinator, Washington University School of Medicine.

285 By early 2016, the DIAN trial had scanned fifty brains: Author's interview with Randy Bateman, January 12, 2016.

285 "The current thinking is you need the cortex loaded with amyloid": Author's interview with Pierre Tariot, January 25, 2016.

285 Ten months after Gail's death, a research team reported: Elizabeth Cohen, "Blood Test Predicts Alzheimer's Disease," CNN, March 9, 2014.

286 In December 2015, the Banner Institute launched GeneMatch: Author's

interviews with Pierre Tariot, January 25, 2016, and Eric Reiman, January 26, 2016.

286 In March 2015, after much badgering by Karla, Jamie DeMoe underwent a week of scans: Visit by Jamie DeMoe to the University of Pittsburgh, March 2015.

287 pharmaceutical company Biogen announced: Gabrielle Strobel, "Biogen Antibody Buoyed by Phase 1 Data and Hungry Investors," Alzheimer's Research Forum, March 25, 2015.

288 Paul Aisen, whose A4 study was about halfway to its goal: Author's interview with Paul Aisen, January 15, 2016.

288 Days later, the Mayo Clinic published findings: Cynthia Koons, "Alzheimer's Debate Revived Even as Biogen's Drug Trial Advances," *Bloomberg News*, March 25, 2015.

288 Reiman recalled his dismay: Author's interview with Eric Reiman, January 26, 2016.

289 As she learned about each new development in the field: Author's interview with Karla Hornstein, 2016.

Index

ABOUT THE AUTHOR

NIKI KAPSAMBELIS's work has appeared in publications around the world, including the *Washington Post*, *Los Angeles Times*, *People*, and the Associated Press. She lives in Pennsylvania.